EXPODACH

Symbolbauwerk zur Weltausstellung Hannover 2000
Roof Structure at the World Exhibition, Hanover, 2000

Herausgegeben von / Edited by Thomas Herzog

Prestel   München · London · New York

Inhalt

Contents

6		Am Projekt Beteiligte / Persons and offices involved in the project
8		Vorworte / Forewords
	Sepp D. Heckmann	Deutsche Messe AG / German trade-fair organization
	Fritz Brickwedde	Deutsche Bundesstiftung Umwelt / Federal German Foundation for the Environment
	Günter Keil	Bundesministerium für Bildung und Forschung / Federal German Ministry for Education and Research
12	Manfred Sack	Das EXPO-Dach am Hermes-See / The EXPO roof at the Hermes Lake
16	Thomas Herzog	Hintergrund und Konzept – Motto und Symbol / Background and conception – motto and symbol
18		Der Architektonische Entwurf / The architectural design
25	Rainer Wittenborn	Die Farben der Hölzer / The colours of the timber
30		Die Membrane / The membrane
32	Thomas Kuckelkorn	Tageslicht-Simulationen / Daylight simulations
34	Christoph Hoffmann, Gerhard Rieger, Andreas Schabel	Weißtannen aus dem Schwarzwald / Silver firs from the Black Forest
38	Martin H. Kessel	Laborversuche: Festigkeit und Steifigkeit der Tannenstämme / Laboratory tests: strength and rigidity of the silver fir stems
40	Julius Natterer, Norbert Burger,	Tragwerksplanung / Structural engineering
46	Alan Müller, Johannes Natterer	Dreidimensionale Rechenmodelle / Three-dimensional calculation models
48		Schalengeometrie / Shell geometry
50	Heinrich Kreuzinger	Dynamisches Verhalten der Dachkonstruktion / Dynamic behaviour of roof construction
52	Jacques-André Hertig	Versuche im Windkanal / Wind-tunnel tests
53	Norbert Burger, Julius Natterer	Innovationen im Holzbau / Innovations in timber construction
		Anmerkungen zum konstruktiven Holzschutz / Notes on constructional means of timber protection
		Brettstapelkonstruktionen / Stacked-plank construction
		Einsatz nicht geregelter Bauweisen / Use of innovative, non-regulated types of construction
57	Werner Kelletshofer, Robert Spengler	Versuche zur Tragfähigkeit / Experimental investigations of load-bearing capacity
58	Martin Pfundt	Arbeitsvorbereitung mit weiter entwickelter CAD/CAM-Software / Advanced CAD/CAM software applications
60	Peter Bertsche	Transport und Montage / Transport and assembly
60		Koppelung der Schirme / Connecting the canopies
61	Martin Speich, Josef Lindemann	Unabhängige Bautechnische Prüfung / Independent constructional controls
63	Gerd Wegener, Bernhard Zimmer	Zur Ökobilanz des EXPO-Daches / Life-cycle assessment of the EXPO roof
66		Daten zur Weißtanne / Data on the silver firs
67		Daten zum Bauwerk / Data on the structure
68		Szenen aus Planung und Werkstatt, vom Bau und von der Eröffnung / Scenes photographed during the planning and in the workshop, during construction and at the opening ceremony
70		Autoren des Buches und beteiligte Institutionen / Authors and institutions contributing to this book
72		Förderer des Bauwerks / Sponsors of the structure
		Impressum / Photonachweis / Photo credits
		Beteiligte Firmen / Companies involved in the project

Am Projekt Beteiligte

Persons and Offices Involved in the Project

Bauherr / Client
Deutsche Messe AG, Hannover
Standort / Location: Messegelände Hannover

Verantwortliches Mitglied des Vorstands /
Representative of the managing board:
Sepp D. Heckmann

Leitung Zentralbereich Technik / Director of Central Technical Office:
Dr.-Ing. Rainar Herbertz

Architekten / Architects
Herzog + Partner BDA, München
Prof. Thomas Herzog, Hanns Jörg Schrade
Projektleitung / Project architect: Roland Schneider
Mitarbeiter / Assistants: Jan Bunje, Peter Gotsch,
Moritz Korn, Thomas Rampp, Stefan Sinning

Durchführung und Abwicklung / Realization
BKSP Projektpartner GmbH, Hannover
Projektleitung / Project supervisor: Ingo Brosch
Mitarbeiter / Assistants: Wilfried Peters, Hans-Joachim Kaub

Tragwerksingenieure / Structural engineers
IEZ Natterer GmbH, Wiesenfelden
Prof. Julius Natterer, Dr.-Ing. Norbert Burger
Mitarbeiter / Assistants: Andreas Behnke, Alan Müller,
Johannes Natterer, Volker Schmidt

Ingenieurbüro Bertsche, Prackenbach
Peter Bertsche,
Mitarbeiter / Assistant: Peter Fitz

Ingenieurbüro kgs, Hildesheim
Prof. Dr.-Ing. Martin H. Kessel, Dirk Gnutzmann
Mitarbeiter / Assistants: Klaus Winkelmann, Georg Klauke

Schwingungsbeurteilung / Vibration report
Technische Universität München, Institut für Tragwerksbau
Prof. Dr.-Ing. Heinrich Kreuzinger

Prüfingenieur / Proof engineers for structural analysis
Ingenieurbüro Speich-Hinkes-Lindemann, Hannover
Prof. Dr.-Ing. Martin Speich, Dipl.-Ing. Josef Lindemann

Farbgestaltung / Colour design
Prof. Rainer Wittenborn, München

Lichtplanung / Lighting design
Ulrike Brandi Licht, Hamburg
Projektleitung / Project supervisors: Mariana Müller-Wiefel, Oliver Ost

Membranplanung / Membrane planning and engineering
IF Jörg Tritthart, Dr.-Ing. Hartmut Ayrle, Reichenau / Konstanz
Engineers and architects for lightweight structures

Projektsteuerung / Project management
Assmann Beraten und Planen GmbH, Hamburg
Dr.-Ing. Wolfgang Henning

Bodengutachten / Soil report
Dr.-Ing. Meihorst & Partner, Hannover

Brandschutz / Fire protection
Hosser, Hass & Partner, Braunschweig

Gründung / Foundations
Renk Horstmann Renk, Hannover

Vermessungstechnik / Surveying services
Drecoll v. Berckefeldt, Hannover

Außenanlagen / External works
Dieter Kienast †, Vogt Partner, Zürich

Vorworte

Forewords

Sepp D. Heckmann

Mitglied des Vorstandes
der Deutschen Messe AG, Hannover,
und verantwortlich für den Bereich Planen und Bauen,
Betrieb und Kommunikationstechnik
der EXPO 2000 Hannover GmbH

Member of the board of the Deutsche Messe AG,
Hanover, and responsible for planning and construction,
operations and communications technology
for EXPO 2000, Hanover

Fritz Brickwedde

Generalsekretär der Deutschen Bundesstiftung Umwelt,
Osnabrück

Secretary-general of the Federal German Foundation
for the Environment, Osnabrück

Messen sind nicht nur nüchterne Geschäftsereignisse. Gefühle, Stimmungen und Erwartungshaltungen beeinflussen hier in ganz starkem Maße alle Kommunikationsprozesse. Die entscheidende Rolle in diesem Umfeld kommt der Messearchitektur zu. Wenn Ausstellungsfazilitäten wie in Hannover Dynamik, Modernität und Individualität ausstrahlen, so beflügelt dies die Stimmungslage aller Beteiligten. Mit dem EXPO-Dach setzt die Deutsche Messe AG für den Messeplatz Hannover, im weiteren aber auch für den Messeplatz Deutschland, einen einmaligen städtebaulichen Akzent. Aufregende Formgebung und eine neuartige Konstruktionsweise machen es nicht nur für Architekturinteressierte zu einem Erlebnis der besonderen Art. Dabei begeistern nicht zuletzt auch die realisierte Holzbautechnik und die ihr zugrunde liegenden Prozesse elektronischer Datenverarbeitung. Die universellen Nutzungsmöglichkeiten des EXPO-Daches, im Freien ausstellen und dennoch beschirmt sein, fügen den hochmodernen Ausstellungsmöglichkeiten des Messeplatzes Hannover eine neue Variante hinzu. Die Pavillons unter dem EXPO-Dach übernehmen dabei optisch eine Scharnierfunktion zu den angrenzenden Messehallen. Das EXPO-Dach, eine einmalige Skulptur in Holz, wird auch nach der Weltausstellung EXPO 2000 als spektakuläres Zeichen für architektonischen Mut und für das Beschreiten neuer Wege Aussteller und Besucher faszinieren.

Trade fairs are not just down-to-earth business events. Emotions, moods and expectations have a major influence on the whole range of communication processes involved here. The decisive role in this context, however, is played by the architecture. When exhibition facilities like those in Hanover communicate such a strong sense of dynamism, modernity and individuality, the elan of all those involved is given a boost. With the EXPO roof, the Deutsche Messe AG (the organization responsible for trade fairs in Hanover) has struck a new and unique urban note – not just for the city, but for Germany as a whole as a venue for trade fairs. Exciting formal design and a new type of construction make the roof an experience of a special kind, and not only for architectural buffs. What makes this structure so fascinating is the timber technology it employs and the electronic data-processing that helped to make it possible. The universal scope afforded by the EXPO roof – creating a place where events can be staged in the open air, yet sheltered from the weather – adds a further dimension to the state-of-the-art trade-fair exhibition facilities in Hanover. The pavilions beneath the EXPO roof also have a visually pivotal function in relation to the nearby trade-fair halls. The EXPO roof is a unique sculpture in timber; and even after the world exhibition EXPO 2000, it will continue to fascinate exhibitors and visitors alike as a spectacular symbol of architectural courage and the will to explore new paths.

Die Deutsche Bundesstiftung Umwelt wurde 1991 als gemeinnützige Stiftung durch ein Gesetz des Deutschen Bundestages ins Leben gerufen. Aus den Zinserträgen eines Stiftungskapitals von nunmehr fast 3,1 Milliarden DM werden jährlich rund 450 Vorhaben mit einem Fördermittelvolumen von ca. 150 Millionen DM in den Bereichen Umwelttechnik, Umweltforschung und Umweltkommunikation unterstützt. Hierdurch sind in der ganzen Bandbreite der Vision einer nachhaltigen Entwicklung Möglichkeiten geschaffen worden, den begonnenen Trend mit umweltrelevanten Projekten anzuschieben. Der Grundgedanke der Stiftung ist nicht die flächendeckende Maßnahme, sondern das beispielgebende innovative Vorhaben.

Die Deutsche Bundesstiftung Umwelt lenkt dabei ihr Augenmerk in erster Linie auf den produktionsintegrierten vorbeugenden Umweltschutz mit einer ganzheitlichen Betrachtung von der Rohstoffauswahl über den Maschinenbau bis zur Prozeßtechnik. Die Innovationshöhe, der Entlastungseffekt für die Umwelt und die Multiplikatorwirkung der Vorhaben unter besonderer Berücksichtigung kleiner und mittlerer Unternehmen sind dabei entscheidend für eine Förderung. Im Bereich Architektur und Bauwesen werden insbesondere interdisziplinäre ganzheitliche Planungen, die Entwicklung ressourcenschonender Bauteile und -produkte sowie herausragende Demonstrationsprojekte gefördert.

Da das EXPO-Dach die Förderkriterien der Deutschen Bundesstiftung Umwelt in besonderer Weise erfüllte, entschloß sich das Kuratorium zu einer dreistufigen Förderung der Planungsphase, der Realisierung und der umfassenden wissenschaftlichen Dokumentation mit einem Gesamtbetrag von 4,1 Millionen DM. Möge die außergewöhnliche Leistung aller Beteiligten weltweit Beachtung und Nachahmer finden.

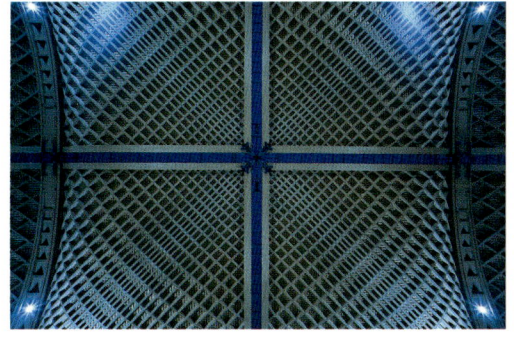
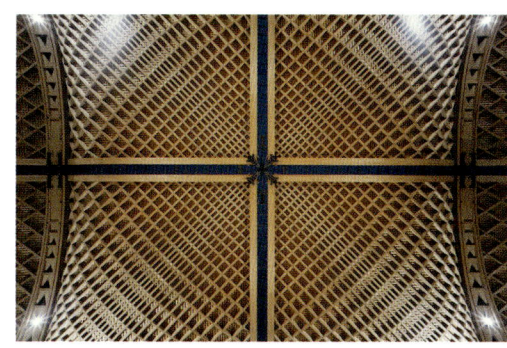

Günter Keil

Bundesministerium für Bildung und Forschung (BMBF), Bonn

Federal German Ministry for Education and Research (BMBF), Bonn

The Federal German Foundation for the Environment was established in 1991 as a non-profit organization by an act of the German Bundestag. From the interest yielded by the foundation's endowment capital – almost DM 3.1 billion – some 450 developments are supported each year in the field of environmental technology, environmental research and environmental communications, with funds amounting to roughly DM 150 million. New scope has thus been created for lasting developments that cover a broad visionary spectrum, and fresh impetus will be lent to an already existing trend towards realizing environmentally relevant projects. The basic goals of the foundation lie in the promotion not of large blanket schemes, but of model, innovative developments.

The focus of the work of the Federal German Foundation for the Environment may be seen in a holistic concept of preventive environmental protection, in which production processes are integrated, with measures ranging from the selection of raw materials to machine construction and processing technology. Decisive factors for granting support are the innovative level of the schemes, the relief they bring to the environment and the multiplying effect they have, especially in the way they take account of smaller and medium-sized firms. In the realm of architecture and construction, the foundation looks for interdisciplinary holistic planning, the development of building elements and products that conserve resources, and outstanding demonstration projects.

Since the EXPO roof met the criteria of the Federal German Foundation for the Environment in a quite special way, the board of trustees decided on a three-stage support measure, covering the planning phase, the construction, and a comprehensive scientific documentation. A total sum of DM 4.1 million was granted for these purposes. May the exceptional achievement of all those involved gain international recognition and followers!

Es war kein Zufall, daß der Plan der Deutschen Messe AG, für die EXPO eine Demonstration der fortschrittlichsten Anwendung des Holz-Ingenieurbaus zu wählen, auf ein auch diesem Thema gewidmeten Förderungsprogramm des BMBF traf: Seit Mai 1988 wurden die bisherigen Förderaktivitäten zum Thema Holz zusammengefaßt; ein neuer Förderschwerpunkt wurde eingerichtet, der sich die Anwendung des Rohstoffs Holz zum Ziel gesetzt hat. Dabei geht es in erster Linie darum, durch technische Innovationen Holz und neue Holzwerkstoffe für alle nur denkbaren Anwendungen geeigneter zu machen und auch völlig neue Anwendungen und Produkte auf der Basis von Holz zu entwickeln.

Seither wurden 20 Millionen DM für neue Projekte investiert.

Der reichlich vorhandene, nachwachsende Rohstoff und Werkstoff Holz hat möglicherweise seine besten Anwendungschancen noch vor sich, wenn es gelingt, das in Deutschland vorhandene technologische Potential für Innovationen in der Holztechnologie zu aktivieren. Eine große Chance liegt dabei in der Nutzung von Technologien und Methoden, die bereits in anderen Industriebereichen, zum Beispiel in hochautomatisierten Fertigungsprozessen, weit entwickelt sind, sowie auch in der Zusammenführung von physikalischen, chemischen und biologischen Technologien für neue Lösungen, z.B. im Holzschutz.

Jede gute Technik benötigt Symbole, die die eigentliche Botschaft – hier: »So beeindruckend kann Holzbau sein« – vermitteln; das EXPO-Dach erfüllt diese Aufgabe beispielhaft.

It was no coincidence that the plan of the Deutsche Messe AG, the German trade-fair organization, to demonstrate the most advanced application of timber engineering in the construction work for the EXPO converged with a support programme of the Federal Ministry for Education and Research (BMBF) that had the same goals. Since May 1988, the various timber support measures implemented by the BMBF have been combined and co-ordinated, and a new focus has been established with the aim of promoting the use of timber as a raw material. First and foremost, this means encouraging technical innovation to make timber and new wood products appropriate for all conceivable applications, as well as developing new uses for wood and new products based on this material.

Since then, some DM 20 million have been invested in new projects.

Wood, as a raw material and building material, is in plentiful supply, and it is also replenishible. Probably the greatest opportunities for its application still lie in the future if one can successfully activate the existing potential for innovation in timber technology in Germany. One area of great promise is the use of technologies and methods that are more advanced in other industrial sectors; for example, highly automated fabrication processes; or the combination of technologies from the realms of physics, chemistry and biology to create new solutions, e.g. in the field of timber protection.

Every good technology needs its own symbols, which should be capable of communicating the intrinsic message – in this case: "Look how impressive timber construction can be!" The EXPO roof admirably rises to the challenge.

Das EXPO-Dach am Hermes-See

The EXPO Roof at the Hermes Lake

Manfred Sack

Ein Haus zu bauen, gehört zu den existentiellen Bedürfnissen des Menschen. Um wieviel elementarer ist es, sich an seine Urform zu halten: an das Dach. Schon unser Sprachgebrauch eröffnet seine Bedeutung. Man errichtet es, um sich gegen die Unbilden der Witterung zu schützen, um »ein Dach überm Kopf zu haben«, damit der Mensch nicht obdachlos, sondern behaust sei; mehr, damit er sich darunter nach Kräften zu Hause fühle. So gehört es zu den alltäglichsten Drohungen, jemandem aus Zorn »aufs Dach zu steigen«, und zu den grausigsten, im »den roten Hahn aufs Dach zu setzen«, ihm also »das Dach überm Kopf anzuzünden«. Das Dach ist Schutz schlechthin – daher auch unser Hang, ihm symbolische Kraft beizumessen. Wundert es, daß Dächer besonderer Art infolgedessen nicht bloß als Bauwerke gewürdigt werden, sondern als Versprechen empfunden, als Mythen?

So geschah es 1972 mit dem »Münchner Dach«, das zum Kennzeichen des ganzen Olympiaparks geworden ist und bis heute das gebaute Symbol für das ehrgeizige Programm der damals ausdrücklich gewünschten »heiteren Spiele« geblieben ist. Es wurde nicht allein für die eleganten Schwünge bewundert, die es vollführt, für seine Transparenz, sondern auch für seine atemraubende Konstruktion, die für möglich zu halten selbst die gewieftesten Köpfe eine Weile gebraucht hatten.

Nun also gibt es das Hannoversche, das EXPO-Dach, unübersehbar das Signet der Weltausstellung 2000 (und künftiger Messen an diesem Ort), verblüffend identisch mit dem Programm, das sich in den drei Worten »Mensch – Natur – Technik« bekanntgibt und Nachhaltigkeit predigt und Innovation. Das zweite meint etwas neu Gedachtes, so noch niemals Dagewesenes; das erste erläutert das Wörterbuch mit »anhaltend, dauernd, lange nachwirkend«. Nachhaltigkeit meint auch den Aufruf, der Welt nur so viel zu entnehmen, wie sie neu zu bilden imstande ist, oder auch: den Respekt des Menschen vor der Natur, wenn er technisch mit ihr wetteifert. Deswegen hat der Architekt Thomas Herzog sich in dergleichen Denkweise von jeher geübt, zusammen mit seiner Mannschaft das EXPO-Dach entworfen und es aus dem reichlich immer neu wachsenden Rohstoff Holz konstruiert und gebaut.

Und nun fliegen dem Bau wie von allein Superlative zu, zum Beispiel der des größten und weitesten Holzdaches der Welt. Zugleich demonstriert es die noch kaum geahnten, geschweige denn ausgeschöpften Chancen, die dem Baustoff Holz innewohnen, die ihm aber vor allem zwischen Donau und Ostsee einfach nicht geglaubt werden: daß Häuser aus Holz so stabil, so haltbar, für vieles ebenso tauglich sind wie Gebäude aus Stein, Beton, Stahl. Dort hält man Häuser aus Holz immer noch für minderwertig, kurzlebig, provisorisch und mutet sie deshalb nur den vermeintlich flüchtigsten und ärmsten Mitmenschen zu, den Asylbewerbern.

Freilich ging es hier, auf dem Hannoverschen Messegelände, am großen rechteckigen Wasserbecken mit dem Hermes-Turm, nicht um das Holz allein und den unaufhörlichen Reichtum seiner Existenz, sondern um die Herausforderung, die alle Dach-, Turm- und Brückenbauer von jeher zu packen pflegt: immer höher und weiter, immer filigraner und eleganter, immer kühner, dabei mit so wenig materiellem Aufwand wie möglich zu bauen, allerdings mit immer schlaueren, listigeren, gewandteren, scharfsinnigeren Konstruktionen. Dieses intelligente Spiel mit dem Holz und die Lust, zu zeigen, daß es mehr kann, als in ihm steckt, indem man es immer raffinierter schneidet, biegt, verleimt, galt nun dem allereinfachsten Zweck, dem ein Bauwerk zu genügen vermag: Menschen Schutz vor Regen, Hagel, Schnee zu bieten, und vor der grellen Sonne auch. Der Schatten, den das Dach wirft, ist dank dem Netzwerk der Gitterschalen und der transluzenten weißen Membrane, mit dem sie bespannt sind, milde. Es ist, als riesele das Licht wie durch einen Filter sanft herab.

Ist Holz das eine Thema, ist Wasser ein anderes. Es wird in den Dachmulden aufgefangen und in Röhren geleitet, die in den gewaltigen vierbeinigen Stützen des Daches angebracht sind, etwa zweieinhalb Meter über der Erde enden und das Wasser in die rechteckigen Wasserbahnen hinabtröpfeln oder -schießen lassen. Diese Bahnen gliedern den überdachten Platz in »Grachten« und in »Inseln« (oder in »Pontons«) und münden in das große Becken nebenan, das nach dem Aussichtsturm »Hermes-See« genannt ist.

Was sieht man zuerst, wenn man sich dem hölzernen Bauwerk nähert? Ein mächtiges, an den Rändern in eigenwilligem Rhythmus schwingendes, von lichthellen Fugen durchzogenes Dach, das von gewaltigen Stützen getragen wird. Sie bestehen aus je vier, fast zwanzig Meter hohen, wie Duckdalben unten sich spreizenden dicken Baumstämmen: zusammen vierzig, über zweihundert Jahre alte Weißtannenstämme aus dem südlichen Schwarzwald. Sie sind, wie alles Holz hier, so gelassen worden, wie sie sind, unbehandelt – nur tiefrote Farbe schmückt und belebt Spalten und Kanten hier und da (die einzige Farbe auch, die man im Holz- und Metallbauwerk eines Konzertflügels findet).

Jeder der zehn Türme, darin die vier dicken Stämme untereinander mit dekorativ zugeschnittenen Holzplatten ausgesteift, trägt also das Dach. Das besteht aus zehn Einzeldächern, deren jedes aus vier Teilen zusammengefügt ist. Schon fangen Assoziationen an, sie metaphorisch begreiflich zu machen. Man denkt an aufgespannte quadratische Schirme, unter die der Sturm gefegt ist und sie umgeklappt hat, aber auch an großblättrige Blüten, die sich weit geöffnet haben.

Wie einfach, wie kompliziert! Allein der Blick von unten auf die zuerst schleierhafte, dann sich als erstaunlich pfiffig zeigende, symmetrisch geformte Blattstruktur dieser zweifach gekrümmten Gitterschalen läßt eine Unmenge elektronischer Rechenoperationen vermuten, mit denen jeder Knoten räumlich haargenau definiert und fixiert ist: um die in

sich selbst stabilen Krümmungen als tragenden Effekt nutzen zu können.

Von weitem wirkt das Dach, als schwappte es an den schnurgeraden Rändern leicht auf und ab, etwas, das dem Bedürfnis nach Regelmäßigkeit und Ebenmaß zu widersprechen scheint. Es kommt nicht leichtfüßig daher, es scheint erst recht nicht zu schweben. Es gibt stattdessen offen zu erkennen, daß es ein stämmiges, sichtlich gewagtes, ein kraftstrotzendes Bauwerk mit wunderlichen filigranen Zügen ist. Tatsächlich erschließt sich seine Wohlgestalt erst auf den zweiten, dritten Blick, in Wirklichkeit über die Kenntnis von der Eigenart der Konstruktion. Sie soll, selbstverständlich, ein ästhetisches Wohlgefühl erregen, vor allem einen möglichst großen Platz überdachen und die Benutzer mit möglichst wenigen Stützen stören.

Es haben darunter erstaunlich viele Menschen Platz, wenn sie welchem Zauber auch immer in seinem Schutze folgen, und vier containerartige Pavillons obendrein. Ach, denkt man zuerst, wie schade, daß sie hier stehen. Dabei sind sie wohlgefällig proportioniert, sind flexibel zu gebrauchen für viele und vieles. Merkwürdigerweise bringen sie Ruhe an den Ort. Das geschieht hauptsächlich durch das Maß und die elementare, klare Gliederung ihrer Fassaden aus Holz und Glas – durch das schöne Bild, das sie, beplankt mit waagerechten Brettern und mit Sperrholztafeln (in einem anderen Ton), und mit deckenhohen Fensterwänden geben.

Davon, daß sich unter dem grandiosen Holzdach ein sorgfältig formulierter Platz befindet, war schon die Rede. Doch Plätze brauchen, um als Räume empfunden zu werden, eine Fassung, sei es durch Gebäude, die sie dicht und fest umgeben, sei es nur durch eine Freitreppe – so wie hier, wo das Terrain leicht abfällt und der Platz dort nun durch sieben, acht helle, mit dem dunklen Asphalt konstrastierenden Steinstufen wie auf ein Podest gehoben wirkt. Doch es erwartet einen dort ja auch etwas nicht Alltägliches.

Building a house is one of the existential needs of man. How much more fundamental it is, then, to confine oneself to the primal form of shelter: the roof! Its significance is revealed in the language we use. Erected to provide protection against the elements, it ensures that one has "a roof over one's head", a place where one feels at home; and many people may "live under the same roof". "The roof's the limit", the highest point; and to make a great commotion is described as "raising the roof". Finally, one of the worst things one can do is to "go through the roof" in anger.

The roof stands for protection in its most basic form; hence our inclination to invest it with symbolic meanings. It is scarcely surprising, then, that a special kind of roof is regarded not just as another building structure, but as something that holds a promise – something mythical.

That was the case with the roof over the Olympic stadium in Munich in 1972, which became the landmark of the entire area. To this day, it has remained the built symbol of an ambitious programme for what were expressly conceived as "serene and happy Games". The roof was admired not just for its elegant curving lines or its transparency, but above all for its breathtaking construction, which even the most astute minds at that time took a while to believe was possible.

Now there is a structure of this kind in Hanover: the EXPO roof, which is quite clearly the symbol and emblem of the World Exhibition 2000 – and of future trade fairs to be held on this site. It reveals an incredible identity with the programme the EXPO announces in its motto "Humankind – Nature – Technology", as well as with the concepts of sustainability and innovation it advocates. Innovation implies something newly thought out, the like of which has never been seen before; while sustainability is defined as "of an enduring nature", something that can be maintained at a certain level or with a lasting effect. The word also contains an appeal to extract only so much from the earth as it is capable of regenerating; in other words, showing respect for nature when we enter into technical competition with it. The Munich architect Thomas Herzog has long been an exponent of this philosophy and, together with his team, he has planned the EXPO roof in this spirit. For that reason, too, he has designed and built it in timber, a raw material whose reserves can be richly replenished.

Suddenly, all manner of superlatives are being heaped on this structure; for example, that it is the biggest, broadest-spanning timber roof in the world. At the same time, it demonstrates the largely unsuspected and scarcely exploited scope offered by timber as a building material. This is something that still arouses disbelief in Germany, from the Danube to the Baltic: the fact that buildings in timber are as stable, as durable, as suitable for a wide range of purposes as structures of stone, concrete or steel. In Germany, timber buildings are still regarded as inferior, short-lived, provisional – good enough for only the poorest and, as one thinks, most transient people in society: those seeking political asylum.

Here, on the Hanover trade-fair site in the south of the city, beside a large rectangular pool of water overlooked by the Hermes Tower, the central concern was not alone timber, of course, despite its infinite richness. It was the challenge felt by all builders of roofs, towers and bridges: the challenge of attaining greater heights and spans with ever bolder, yet more slender and elegant forms of construction; building with a minimum of materials; building with ever more refined, ingenious, skilful, perceptive forms of construction. This intelligent game played with timber, plus the desire to show that there is more to this material than one suspects – by cutting, bending and gluing it in ever more subtle forms – serves the simplest purpose a structure can fulfil: protecting people against

rain, hail and snow and against the glaring sun. Thanks to the grid-like pattern of the lattice shell structures and the translucent white membrane with which they are covered, the roof casts a mild shadow. Light seems to float down softly as if through a filter.

If timber is one of the major themes of this project, water is another. Collected in the hollows of the roof, it is funnelled into pipes fixed vertically in the huge, four-legged masts that support the roof. The pipes terminate roughly two and a half metres above the ground, allowing the water to trickle or pour into the rectilinear grid of water channels beneath. These strips of water or "grachts" articulate the covered outdoor space into a series of islands or "pontoons" and finally flow into the large adjoining Hermes Lake, which takes its name from the viewing tower standing in the middle.

On approaching the new timber structure, what does one see first? A powerful roof, curving at the edges in a distinct rhythm, dissected by joints through which light shines, and supported by massive columns. These consist of four roughly 20-metre-high stout tree stems, splayed apart like mooring piers. In all, there are 40 of these stems, cut from more than 200-year-old fir trees in the southern Black Forest. Like all the timber used here, they have been left in a natural, untreated state, with only the occasional split and arris and their deep red colour as decoration (the only colour, incidentally, to be found in the wood and metal construction of a grand piano).

Each of the ten masts, consisting of four thick stems tied together and braced by wood sheets cut out in decorative form, bears a self-contained section of the roof. The whole structure comprises ten canopies, each of which is made up of four segments. Associative images come to mind that make the roof comprehensible in a metaphorical form. One thinks of a series of square stretched umbrellas, beneath which a gale has swept and turned them inside out; or one is reminded of flowers with gigantic, fully opened petals.

How simple, yet how complex! Alone the view from below, up to what at first seems a diaphanous, then an astonishingly lively, symmetrically laid-out leaf-like structure in the form of double-curved lattice shells, suggests a vast amount of electronic calculating operations, by means of which each of the intersections was precisely defined and three-dimensionally located to exploit the innate structural stability of the curve.

From a distance, the roof looks as if it would wash lightly up and down in wave-like form along its straight edges – a feature that seems to contradict the need for regularity and evenness. The roof does not trip lightfootedly. It does not seem to float at all. On the contrary, it openly declares itself to be a sturdy, visibly daring, muscular structure with strangely filigree features. The subtleties of its form become evident only at second or even third sight, and really only through a knowledge of the unique nature of the construction. Aesthetically, of course, it is meant to create a sense of well-being; but above all, it has to cover as large an area as possible, while causing users a minimum of inconvenience through intermediate columns.

There is room for an incredible number of people beneath this protecting roof – to whatever magic they are drawn. There are also four container-like pavilions. "Ah, what a pity they stand here!" one thinks at first. But the pavilions have pleasing proportions; they are flexible in use and can accommodate a large number of people and functions. Curiously enough, they also lend this place a sense of peace and calm. This is largely the result of their scale, the elementary and incredibly clear articulation of their wood-and-glass facades, and their fine appearance. They are clad with horizontal boarding and plywood panels of contrasting tone, and opened up with room-height glazing.

One spoke of a carefully formulated public space beneath this grand timber roof; but to create a sense of space, locations need some means of demarcation. This may be in the form of buildings that provide a tight enclosure, or a flight of steps, as is the case here, where the site slopes down gently and the square appears to be elevated like a platform with seven or eight light-coloured stone steps contrasted with the dark asphalt. Something quite out of the ordinary awaits the visitor here.

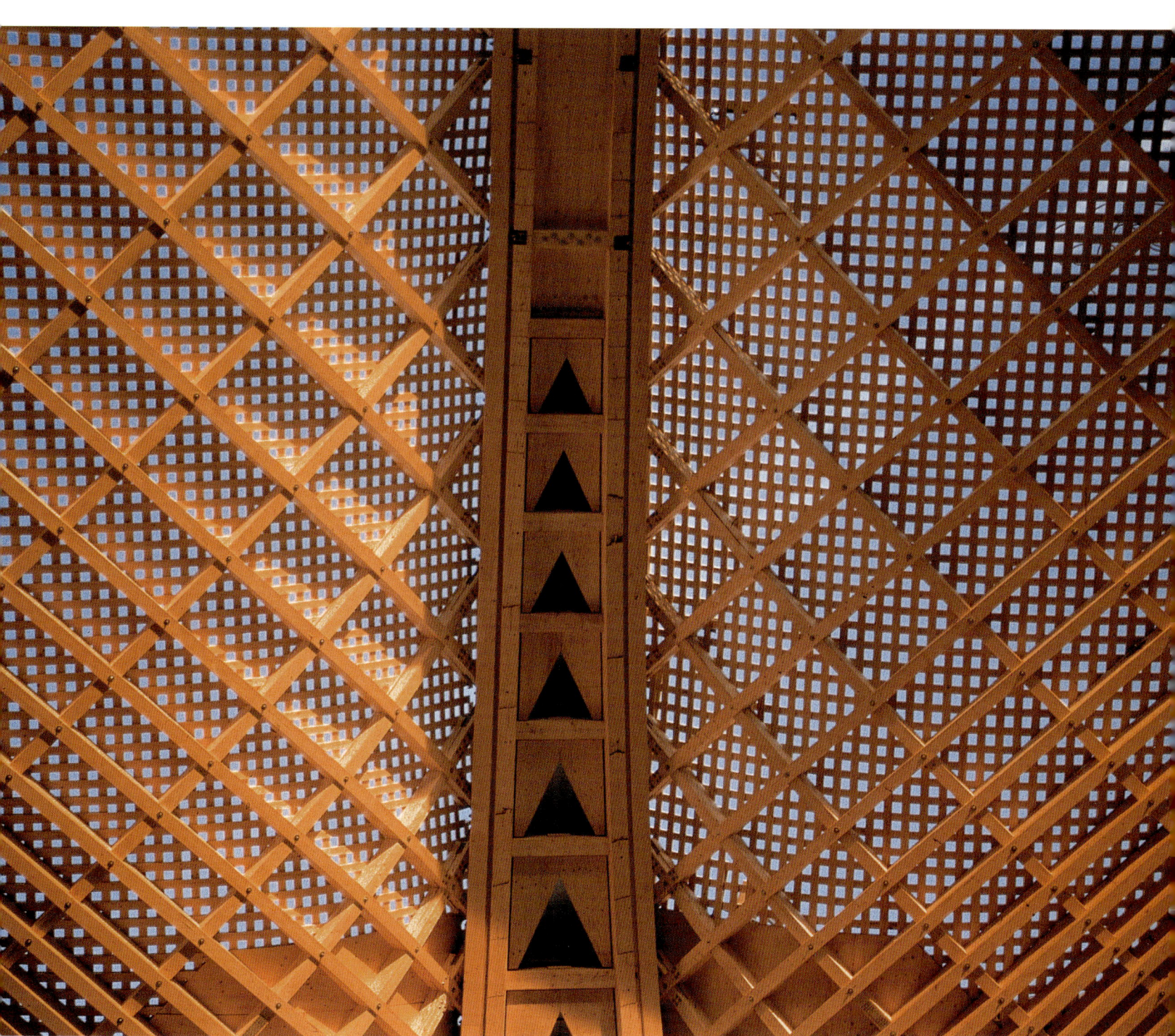

**Hintergrund und Konzeption –
Motto und Symbol**

**Background and Conception –
Motto and Symbol**

Thomas Herzog

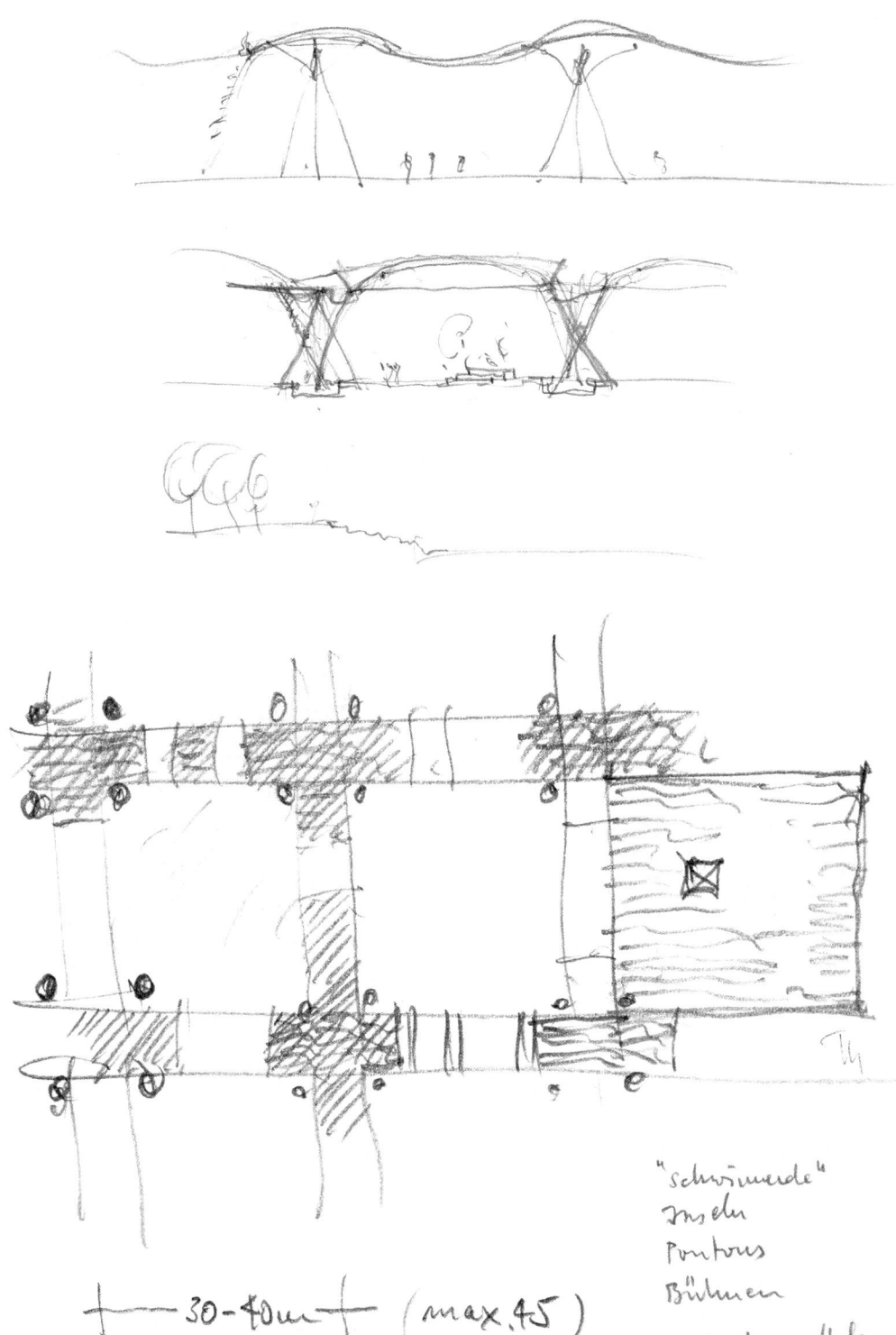

Erste Entwurfsskizzen des Architekten / Architect's initial design sketches

Will man das Motto: »Mensch – Natur – Technik« mit einem Bauwerk eindrucksvoll und überzeugend darstellen, so müssen die drei Aspekte in ihrem Zusammenwirken in ein prägnantes Zeichen umgesetzt sein.

Das Dach ist die Urform des Schutzes, den die Menschen seit jeher gegen die Unbill der Witterung bauen. Es ist für die EXPO 2000 deshalb ein geeignetes architektonisches Motiv, da Freiflächen unter den klimatischen Bedingungen Hannovers auf dem Ausstellungsgelände bis heute wetterabhängig und daher nur eingeschränkt zu nutzen sind. Andererseits würde eine Beschränkung der Präsentationen von Exponaten und Aktionen auf den Innenbereich der vorhandenen Messehallen einen erheblichen Verzicht auf Erlebnismöglichkeiten für die Besucher auf dem Gelände der Weltausstellung bedeuten.

Ein großes, weites, gleichermaßen kraftvoll und elegant wirkendes Dach in zentraler Lage des Geländes definiert und schützt einen Ort, wo Musikdarbietungen, Shows jeder Art, künstlerische Aktionen und vieles mehr an Aufführungen vor großem Publikum wettergeschützt und dennoch im Freien stattfinden. Dort stehen auch einzelne Pavillons unterschiedlicher Nutzung, wo während der EXPO unter anderem Restaurants zum Verweilen einladen. Bei schlechtem Wetter sind Akteure, Veranstalter und Besucher geschützt; so müssen Veranstaltungen nicht wegen eines Regens abgesagt werden. Brennt die Sommersonne, so bietet das Dach angenehmen Halbschatten.

Durch das Zusammenwirken von transparenten und geschlossenen, linearen und flächigen Teilen des Daches entsteht eine bauliche Gesamtstruktur, welche die Einsatzmöglichkeit von Holz als nachwachsendem Rohstoff in eindrucksvoller Dimension und neuartiger Form darstellt. Man leistet in dieser Weise einen unmittelbar verständlichen Beitrag zum Begriff der »Natur«, der sich im Motto der EXPO 2000 findet.

Komplementäres Material ist Wasser. Durch die Art der Ausbildung der Randflächen entsteht der Eindruck, daß das riesige Dach auf einer entsprechend großen Wasserfläche steht, wobei aufgrund des modularen Aufbaus diese Bodenbereiche auf vielfältige Weise differenziert gestaltet werden können.

Die geometrisch klar definierte Dachkonstruktion findet ihre Entsprechung in der geometrischen Ordnung am Boden.

Der Aspekt »Technik« wird betont, indem der letzte Stand des Ingenieurwissens bei der Ausbildung des Tragwerks und der Membranhaut des Daches zur Wirkung kommt. So entstand durch den Einsatz von modernen Fertigungsverfahren in Verbindung mit handwerklichem Können ein Großbauwerk als High-tech-Gebilde aus Holz, wettergeschützt durch eine lichtdurchlässige Membrane, das als architektonisches Symbol für das Motto der Weltausstellung steht.

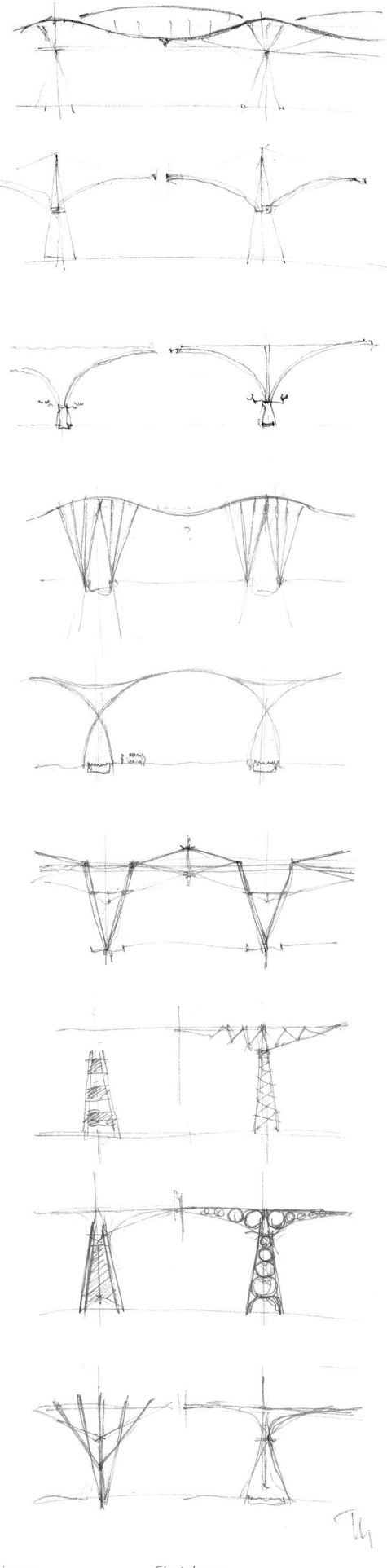

If one seeks to represent the motto of the world exposition, "Humankind – Nature – Technology", in an impressive, convincing built form, the three individual aspects have to be translated into a striking image, taking into account the effects they have in combination with other.

The roof is the primal form of protection, which man has built since time immemorial to shield himself against the elements. For EXPO 2000, therefore, it is a most appropriate architectural motif; for in the climatic conditions prevailing in Hanover, the use of the open areas on the exhibition site is dependent on the weather and would be possible for only limited periods. On the other hand, restricting the presentation of exhibits and activities to the interiors of the existing halls would mean a considerable sacrifice in terms of what visitors to the EXPO site could see and experience.

An extensive roof structure, powerful and at the same time elegant, has, therefore, been erected in a central position on the site to define and protect a place where musical performances, shows of all kinds, artistic activities and many other events can be presented to a large audience in the open air, yet sheltered from the vagaries of the weather. Also situated here are a number of pavilions with various uses. During the EXPO, restaurants and other facilities will invite visitors to sit down for a while and relax. When the weather is bad, performers, organizing firms and the public can rely on the protection of the roof, and the programme of events will not have to be cancelled because of rain. When the summer sun is at its hottest, the roof also provides pleasant half-shaded areas.

Through a combination of transparent and opaque surfaces, linear and planar elements, a universal roof structure was created that demonstrates the scope for using timber, as a replenishable raw material, on an impressive scale and in an innovative form. The structure thus makes an immediately comprehensible contribution to the aspect of "nature", contained in the EXPO 2000 motto.

Water was used as a complementary element. The design of the peripheral areas creates the impression that the huge roof structure is erected over an equally large area of water, although the modular form of construction allows the ground beneath to be designed and laid out in a wide variety of ways.

The geometric, clearly articulated structure of the roof is reflected in the grid-like order of the ground at the base.

The aspect of "technology" described in the EXPO motto is demonstrated by the application of state-of-the-art engineering science in the design of the load-bearing structure and the membrane skin of the roof. In this way, through the use of modern methods of prefabrication in combination with trade skills, a large-scale, high-tech timber structure was created, protected against the weather by a translucent membrane. The roof is, therefore, an architectural symbol and embodiment of the world exhibition motto.

Skizzen zu verschiedenen Formen des Tragwerks

Sketches: various forms of load-bearing structure

Der Architektonische Entwurf

The Architectural Design

Thomas Herzog

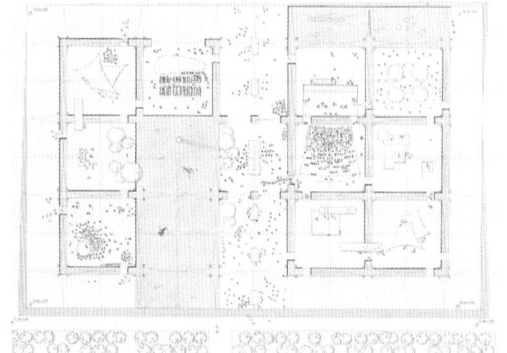

Vorentwurf:
Grundriß mit Wasseranlage (»Grachten«)

Preliminary design:
plan showing grid of water channels ("grachts")

An der östlichen und westlichen Peripherie des Geländes der Weltausstellung stehen die Neubauten der Nationen-Pavillons. Im Mittelbereich des Geländes, zwischen der »Allee der Vereinigten Bäume« und den Südseiten der Hallen 25 und 26, ist eine großräumige architektonische Komposition verwirklicht, bestehend aus neu angelegten Wasserflächen, Ausstellungspavillons und einem großen, aus Schirmen gebildetem Dach. Diese Schirme bilden einzeln stehende riesige Skulpturen, deren Dachfläche jeweils zu ihrer Mitte hin entwässert wird. So entstand an zentraler Stelle des Messegeländes ein gegen Niederschläge geschützter Freiraum, der als solcher auf Dauer erhalten bleiben soll.

Grundgeometrie

Das Großdach besteht aus einzeln stehenden Schirmen mit Dachflächen von circa 40 m x 40 m Seitenlänge, also zusammen 16 000 m² in über 20 m Höhe. Die Bodenflächen sind geometrisch durch »Pontons« (optisch schwimmende Flächen) und »Grachten« (circa 5 m breite Wasserstreifen) definiert und in differenzierter geometrischer Überlagerung mit den Dachelementen gestaltet.

Auch die Entwässerung des Großdachs findet als sichtbar gemachter Vorgang im Zentrum von jeweils vier Pylonen in der Mitte der einzelnen Schirme statt.

Integrale Planung

Nach Konzeption und Entwurf des Architekten, die alle Elemente bereits einschlossen, wurde die Form des Tragwerks in einem Prozeß integraler Planung und Entwicklung mit den Ingenieuren im Detail erarbeitet – wobei Phasen mit Modellbau, Berechnung, Simulation, Windkanalversuchen, Belastungstests und Designstudien über Monate hin immer wieder neu durchlaufen wurden.

Tragkonstruktion

Leichte Flächen aus doppelt gekrümmten Gitterschalen überführen ihre Kräfte in zentrale, mächtige Tragstrukturen aus Holzstämmen. Architektonisches Ziel ist es, die Intelligenz der gewählten Prinzipien und die Schönheit der gefundenen Form in einer modernen technischen Großstruktur zur Wirkung zu bringen. Material, Ideen und Ausführung kommen aus Deutschland als dem Gastgeberland der EXPO 2000; sie zeigen ein markantes Exponat heutigen baulichen Könnens auf der Weltausstellung.

Material und Qualitätssteuerung

Zentral stehende Vollholzstützen mit allseits auskragendem, sie schützendem Dach entsprechen dem Prinzip des konstruktiven Holzschutzes von vornherein in günstiger Weise. Eine maßgebliche Rolle spielt die laufende Qualitäts- und Feuchtekontrolle der Stämme, die bereits an den noch ungefällten Bäumen durch Ultraschallmessungen im Forst beginnt. Derartige Meßmethoden sind heute in Europa noch sehr selten und bedürfen der Zustimmung im Einzelfall. Die Weißtanne-Vollholzstützen stammen aus dem südlichen Schwarzwald; die Forstverwaltungen sind besonders daran interessiert, diese bis zu 50 Meter hohen Bäume auszulichten, um neuem Wachstum Platz zu schaffen. So verjüngt sich ein Wald ohne Kahlschlag auf kleinster Fläche. Je mehr Holz in Bauwerken verwendet wird, umso besser ist dies für die Zukunftschancen unserer Wälder.

Aussteifung

Naturgemäß benötigt ein Bauwerk dieser Dimensionen wirksame Windaussteifungen. Bei Türmen und Masten werden diese im Holzbau heute zumeist in Stahl ausgeführt. Hier aber sind diese Aussteifungsverbände durch großflächige Holzverbundplatten ersetzt. Architekten und Ingenieure haben auch zu definieren, wo Vollholz, wo Leimholz und wo Furnierholz (jeweils unter optimaler Ausnutzung der spezifischen Eigenschaften nach konstruktiven, verarbeitungstechnischen und umweltrelevanten Kriterien) eingesetzt wird. Dies ist auch unter Gesichtspunkten von Recyclieren und späterem Trennen und Wiederverwenden verschiedener Materialien wichtig.

Dachhaut

Die lichtdurchlässige Dachhaut besteht aus nicht brennbarem Material, dessen perfekte Oberfläche hohe Selbstreinigungswirkung hat. Das Erfassen der räumlichen Geometrie, der Zuschnitt der Bahnen mit dreidimensionalem Verlauf der Stöße und die Detailausbildung sowie das Verlegen der Membranflächen auf den geometrisch komplexen, gegensinnig doppelt gekrümmten Dachflächen ist eine anspruchsvolle Aufgabe. Die Helligkeitsverhältnisse unter der großen Dachfläche sind mit neuen Tageslicht-Simulationsprogrammen ermittelt und über die Farbgebung mit beeinflußt.

Wasser

Die Fassung am Rand der Freifläche erweckt den Eindruck, daß das riesige Dach auf einer großen, durchgehenden Wasserfläche steht. Die modulare Einteilung des Bodens erlaubt die Ausbildung der bandartigen Wasserkanäle. Entwässert wird das Großdach über abgehängte Rohre, deren unteres Ende so ausgestaltet ist, daß das Regenwasser in Augenhöhe der Besucher sichtbar geführt wird, im Kreuzungspunkt der »Grachten« landet und Regen, vor dem man sich üblicherweise nur schützt, hier als schönes, vital wirkendes Naturelement erlebbar wird.

Ästhetischer Ausdruck

Die farblich zurückhaltende, aber differenzierte Großstruktur macht nachdrücklich auf moderne Gestaltungsmöglichkeiten von Holzkonstruktionen aufmerksam. Sie schärft die optische Wirkung, unterstützt die komplexe räumliche Geometrie der Bauteile und erhöht so die Lesbarkeit des architektonischen Entwurfs.

Vorentwurf: Ansicht bei Tag / Preliminary design: view of roof by day

Vorentwurf: Ansicht bei Nacht / Preliminary design: view of roof at night

Für die architektonische Grundkonzeption ist von Bedeutung, daß es hier nicht primär darum geht, unter Verwendung von Holz eine möglichst leicht und transparent, fast ›entmaterialisiert‹ wirkende Dachkonstruktion zu realisieren. Solche Vorstellungen sind richtig, wenn es sich um künstlich hergestellte Werkstoffe handelt; deren hohes spezifisches Gewicht in Verbindung mit der Bemühung um große Effizienz bei der Ausbildung des Tragsystems legt eine maximal mögliche Querschnittsreduzierung nahe, wie dies bei Metallkonstruktionen anzustreben ist. Im Fall des EXPO-Daches geht es jedoch auch darum zu zeigen, welche Ausdrucksfähigkeit in modernen Tragwerkskonstruktionen mit dem Universalbaustoff Holz erreichbar ist; man akzeptiert gerne, ja wünscht sogar, daß die stoffliche Materialwirkung ästhetisch stark präsent ist und nicht eine primär filigrane Wirkung gestaltbestimmend wird.

Als besondere Aspekte standen für den Architekten im Vordergrund:
- Die Möglichkeit, in Deutschland verfügbares Massivholz einzusetzen und Rippenschalen aus vernagelten und nur partiell verleimten Brettlagen mit einer differenzierten optischen Wirkung auch im Rahmen eines Großbauwerks auszubilden.
- Bei Entwurf, Planen, Berechnen und Fertigen in geometrisch äußerst komplexen Strukturen den Einsatz von Holz und Holz-

Lage des EXPO-Daches (gebaute Form) im Ausstellungsgelände

Location of EXPO roof (built form) in exhibition grounds

werkstoffen aller Art im Hochbau in schöner, großzügiger Form, auf neuartige Weise zu demonstrieren.
- Die strukturelle und ästhetische Balance zu erreichen zwischen der Leichtigkeit des Materials Holz einerseits und andererseits der starken Dimensionierung der Bauteile; sie ist angesichts der den Berechnungen zugrunde gelegten maximalen Schneehöhen und Windbelastungen gefordert.

Ein zeitgemäßer Holzbau

Konstruktion und Gestaltung des Großdaches führen unterschiedliche Möglichkeiten eines zeitgemäßen Holzbaus vor Augen:

Die bei den einzelnen Schirmen zentrisch stehenden hölzernen Masten sind gegen Verformen in horizontaler Richtung mit großflächigen, durch Rippen stabilisierte Kertoflächen ausgesteift, die in die Vollholzstämme auf die Höhe von 16 m durchgehend einbinden. Auf diesen Masten steht ein stählernes Kernstück – dort wo alle Kräfte kulminieren. Von dort kragen große, in den Untergurten gekrümmte und aus Brettschichtholz hergestellte Träger aus, die als Viergurtkonstruktion mit geneigten Seitenflächen ausgebildet sind. Sie sind gegen seitliche Verformungen mittels eines K-Verbandes in der Ebene der Untergurte ausgesteift.

Zwischen diesen Kragträgern – außen auf ihren Untergurten gelagert – befinden sich doppelt gekrümmte Gitterschalen mit kontinuierlich über die Diagonale von innen nach außen abnehmenden Krümmungen. Sie bestehen aus Rippen in Brettstapelkonstruktion mit unterschiedlichen, dem Kräfteverlauf entsprechenden Abständen. So ist der Kraftfluß stark gestaltbestimmend und als architektonisches Motiv ablesbar.

Das Tragwerk ist für einen Holzbau neuartig. Doppelt gekrümmte Gitterschalen als Brettstapelkonstruktion wurden bisher, erstmals vor wenigen Jahren, nur an kleineren Prototypen erprobt. Das Prinzip ist äußerst effizient und kommt hier erstmals in Form von Sattelflächen großer Dimension zur Anwendung.

Das Erfassen der Kräfte, der Schnittgrößen und die statischen Spannungs- und Verformungsnachweise stellen an das fachliche Wissen und die Kompetenz der Ingenieurbüros äußerste Anforderungen. Erst umfangreiche Versuche im Windkanal brachten wichtige Entscheidungshilfen. Auch die Kapazität der eingesetzten elektronischen Rechner lag am Leistungslimit.

Verarbeitung und Herstellung folgte durch eine Gruppe mittelständischer Unternehmen, welche sich die Baugruppen: Turm, Kragträger und Gitterschalen aufteilten. Im Gegensatz zu den anderen Bauteilen sind die Schalen in nur geringem Maße vorgefertigt. Sie wurden in sieben großen Lehrgerüsten unmittelbar neben der Baustelle produziert.

Die einzelnen Baugruppen sind durch CNC-gesteuerte Maschinen bearbeitet. Die im Rechner erfaßte und dargestellte räumliche Geometrie wurde in die Fertigungsmaschinen übertragen und sorgt für die erforderliche Präzision. Allein die komplexe Form der im Querschnitt sich kontinuierlich verändernden Fläche der Rippenschale bedingt die Definition laufend unterschiedlicher räumlicher Koordinaten über die gesamte Dachfläche.

Es ist offenkundig, daß mittelständische holzverarbeitende Familienbetriebe in der Lage sind, durch sinnvolle Arbeitsteilung und Koordination der Aktivitäten untereinander auch neuartige und ambitionierte Großbauten gemeinsam unter Einsatz des letzten Standes der Technik zu realisieren. Die bauliche Großstruktur des Holzdaches zeigt sowohl für die Besucher der Weltausstellung als auch für künftige Messebesucher die faszinierenden Möglichkeiten moderner Holzkonstruktionen auf.

Flanked along the eastern and western edges by the new national pavilions, a spacious architectural composition has been realized at the centre of the site between the "Avenue of United Trees" and the southern faces of halls 25 and 26. It consists of newly created areas of water, exhibition pavilions and the large roof structure in the form of a series of canopies. The canopies resemble huge free-standing sculptures, the roof areas of which are drained at the centres. In this way a large outdoor space was created at the heart of the trade-fair site, protected against the rain and designed to remain in this location permanently.

Basic geometry

The large roof structure consists of ten individual canopy elements each roughly 40 x 40 m in size. It thus covers a total area of 16,000 m² at a height of more than 20 m. The ground beneath is geometrically articulated by a series of artificial islands or "pontoons" and roughly 5-metre-wide strips of water or "grachts". The ground area is laid out to counterpoint and complement the geometry of the roof.

The drainage of this extensive structure is also designed in such a way that it forms a visual attraction between the four columns of the supporting towers at the centres of the individual canopies.

Integral planning

In accordance with the architects' concept and design, which included the entire range of elements, the form of the load-bearing structure was elaborated in detail in a process of integral planning and development carried out in collaboration with the relevant engineers. As part of this process, various phases such as model-building, calculations, simulations, wind-tunnel trials, loading tests and design studies were repeated over a period of many months to explore alternative forms.

◁ Modell des Vorentwurfs
▷ Entwurfsmodell mit Pavillons

◁ Preliminary design model
▷ Design model with pavilions

Load-bearing construction

The loads from the lightweight double-curved lattice shells are transmitted to powerful central structures assembled from the trunks of trees. The architectural goal was to combine the intelligence of the agreed principles of construction with the beauty of the form derived from this to create an effective, modern, large-scale technical structure. The idea, the materials and the execution are all from Germany, the host country of EXPO 2000. Together, they provide a striking example of modern constructional skills at the world exhibition.

Materials and quality control

The structural form, comprising central columns in solid timber with sheltering roof areas cantilevered out on all sides, is a good example of the principle of providing timber protection by constructional means.

The constant quality and moisture controls to which the tree stems were subject played a crucial role in this project and began in the forest, where ultrasonic tests were carried out before the trees were felled. Methods of measurement of this kind are still rare in Europe and have to be justified on the merits of each individual case. The solid silver-fir stems for the columns come from the southern Black Forest. Forestry offices are particularly interested in the kind of thinning-out process implemented in conjunction with this scheme, in which trees of up to 50 metres in height were removed from the forest, making room for new growth. In this way, the forest can be rejuvenated without clear-cutting whole areas. The more timber that is used in buildings, the better the future of our forests becomes.

Bracing

A structure of these dimensions inevitably requires effective wind bracing. This will usually be in steel where the towers and masts are constructed in timber. In the present project, however, the bracing members consist of large-area composite timber sheeting. The architects and engineers also had to decide where solid timber members, glued laminated timbers and laminated wood sheeting could be used most effectively, optimally exploiting the specific characteristics of these materials according to structural, processing and environmental criteria. Considerations such as these also play an important role in the context of recycling – in the subsequent dismantling and reuse of the various materials.

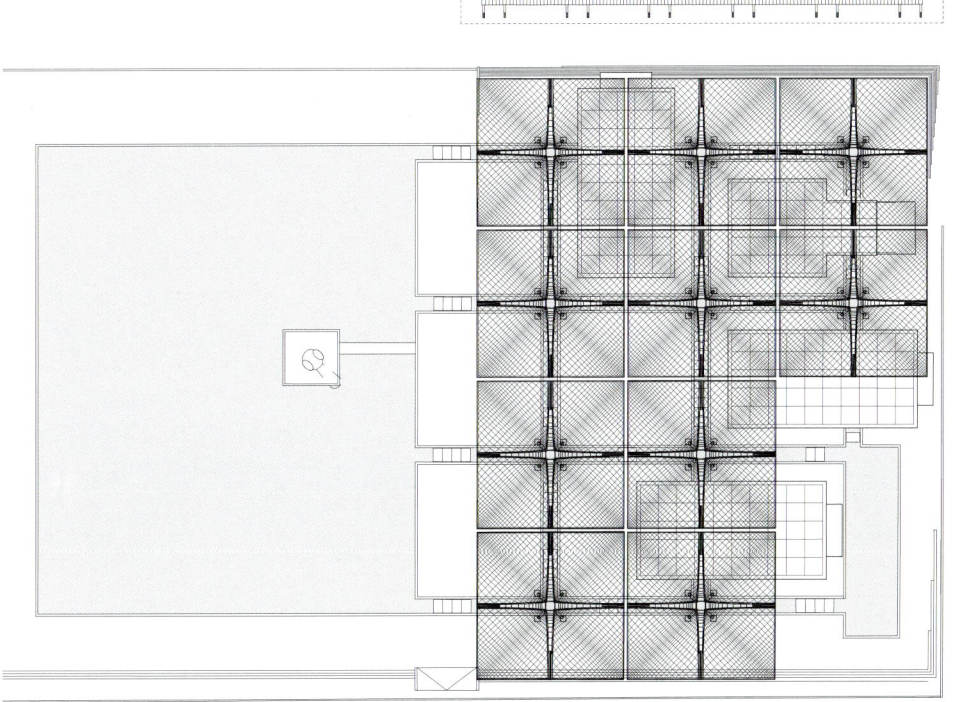

Dachaufsicht mit Grundriß
Maßstab 1:2000

View of roof from above with layout plan
scale 1:2000

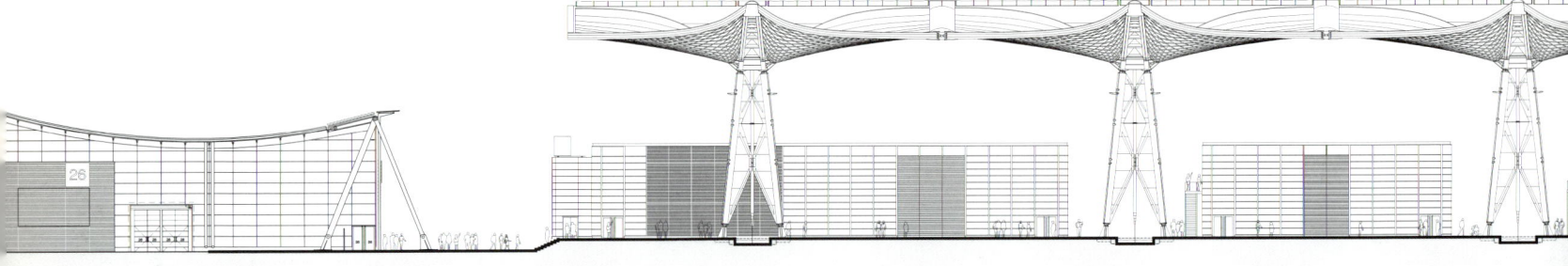

Ansicht von Westen   Maßstab 1:750 / West elevation   scale 1:750

Roof skin

The translucent roof skin consists of a non-combustible material, the perfectly smooth surface of which possesses a high self-cleansing quality. Comprehending the spatial geometry and the three-dimensional jointing of the membrane, cutting and laying the individual sections, and designing the details for the complex double- and counter-curving roof surfaces posed a challenging task. The lighting conditions beneath this enormous structure were calculated with the use of new daylight-simulation programs and modified by means of the coloration.

Water

The enclosing strips of water at the edge of the open space create the impression that this huge roof has been raised over a large, continuous stretch of water. The modular division of the ground area permitted the creation of strip-like canals. The large roof is drained via suspended pipes, the lower ends of which are designed so that the rainwater pours out at eye level as a visual attraction and can be seen by visitors as it falls into the intersections of the canals or "grachts". As a result, rainwater, which is usually something one avoids, is experienced here as a lovely, vital natural element.

Aesthetic expression

This large-scale structure, restrained yet varied in its coloration, is a clear demonstration of the scope for modern design afforded by timber construction. It allows the creation of intense visual effects and a complex spatial geometry within the various elements, and thus increases the legibility of the design.

An important aspect of the architectural concept was that, in opting for a timber structure, the design was not primarily concerned with achieving a maximum degree of lightness and transparency – an almost "dematerialized" form of roof construction. Concepts of this kind may be appropriate for artificially produced or synthetic materials, the high relative density of which, coupled with the need for efficiency in the design of the load-bearing system, makes a minimization of the cross-sectional dimensions desirable. Metal forms of construction are an example of this. In the case of the EXPO roof, however, the aim was also to demonstrate the expressive capacity of timber as a universal building material when used in modern load-bearing structures. In other words, it was regarded not just as acceptable, but desirable, to demonstrate the material qualities of the construction in an aesthetic form rather than achieving a slender, tracery-like effect.

The following aspects played a special role in the design:
- the possibility of using large, solid timber members that were available in Germany; and the production of ribbed lattice shells, consisting of nailed and only partially glued stacked planks, to achieve a varied visual effect even in such a large-scale structure;
- demonstrating in the various stages of the scheme (design, planning, calculation and manufacturing) how timber and composite wood products can be used in an innovative manner in all kinds of building structures – even where the geometry is extremely complex – to create elegant, ample forms;
- achieving a structural and aesthetic balance between the lightness of timber on the one hand, and the bold dimensions of the constructional elements on the other – dimensions that are required to resist the maximum calculated snow and wind loads.

A modern timber structure

In its construction and design, the roof conveys an impression of the wide scope that exists for the modern use of timber.

The masts at the centres of the individual canopies are braced against horizontal deformation by large-area Kerto laminated wood sheets with rib stiffening. The sheets are

Arbeitsmodell: Lehre für die Gitterschale

Working model: centring for lattice shell

Modell Gitterschale

Model of lattice shell

Konstruktionsmodell Einzelschirm

Construction model: single canopy

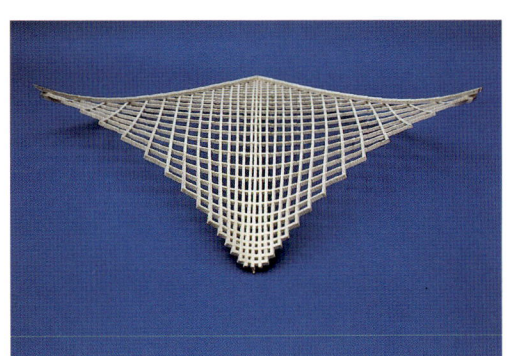

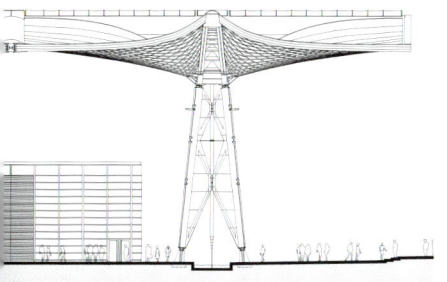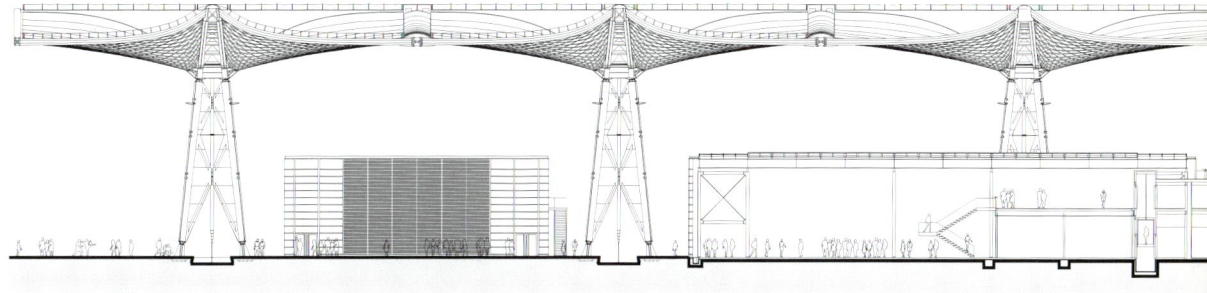

Schnitt mit Ansicht von Süden   Maßstab 1:750 / Section and south elevation   scale 1:750

fixed continuously into the solid timber tree stems up to a height of 16 metres. On top of each of the masts, at the point where all loads culminate, is a steel core element, from which large trusses with curved lower chords cantilever out. The four-part trusses, with inclined side faces, are constructed of laminated glued timbers and are braced against lateral deformation by K-shaped members in the plane of the lower chords.

Between these cantilevered trusses, and supported externally on their lower chords, are double-curved lattice shells, the curvature of which diminishes continuously along the diagonal from inside to outside. The ribs of the lattice shells, consisting of stacked planks, are laid out at spacings that vary according to the loads to be borne. The transmission of loads, therefore, had a great influence on the form of the roof and is legible as an architectural motif.

Building this type of structure in timber is a new development. Double-curved lattice shells in stacked-plank construction were first tested a few years ago and have been used only in smaller prototype projects. The structural principle is extremely efficient and is used here for the first time in conjunction with large-scale saddle-shaped elements. Determining the loads to be borne, specifying the dimensions of the members and calculating the potential stresses and deformation made enormous demands on the professional knowledge and competence of the engineering offices collaborating on this scheme. Extensive wind-tunnel tests were required to obtain the information needed in the decision-making process. The computers used were also taxed almost to the limits of their capacity.

The processing, production and erection of the elements were carried out by a group of medium-sized companies, which divided the construction work into three sections: masts, cantilevered girders and lattice shells. In contrast to the other elements, the construction of the lattice shells involved only a small amount of prefabrication. They were assembled on seven large centring structures immediately adjoining the site.

Machines coupled to computerized numeric control (CNC) systems were used to produce the individual structural elements. The three-dimensional geometry calculated and depicted by the computer was transferred to the production machines and ensured the necessary precision. Alone the complex form of the ribbed lattice shell, with a curvature that constantly changes in cross-section, had to be defined by a large number of different spatial co-ordinates over the entire area of the roof.

The scheme shows that, with a sensible division of responsibilities and the co-ordination of activities, it is possible for medium-sized, family-owned timber processing and construction firms to collaborate

Entwurf: Einzelschirm /
Design stage: view of single canopy

with each other to build ambitious, innovative, large-scale structures, employing state-of-the-art technology. For visitors to the world exhibition and to future trade fairs in this city, this extensive roof structure will be a lasting demonstration of the fascinating scope afforded by modern forms of timber construction.

Konstruktionsvarianten für Stahlfuß und Stahlpyramide

Alternative forms of construction for steel feet and steel pyramid

Modelle: Varianten zur Turmaussteifung

Models: alternative forms of tower bracing

Modell des Stahlfußes

Model showing steel feet

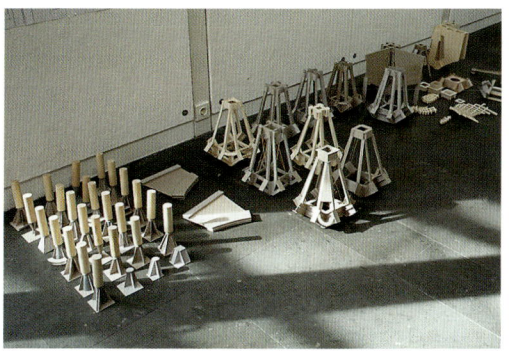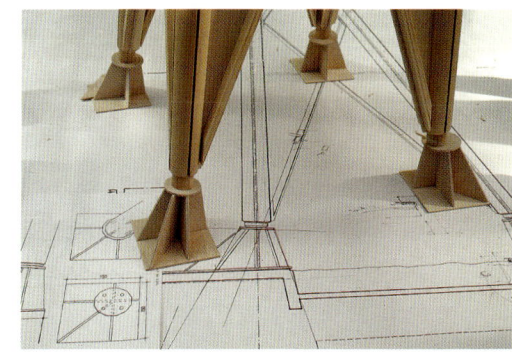

Dachaufsicht und Ansicht Holzkonstruktion
Maßstab 1: 300

Top view of canopy and elevation of timber structure
scale 1:300

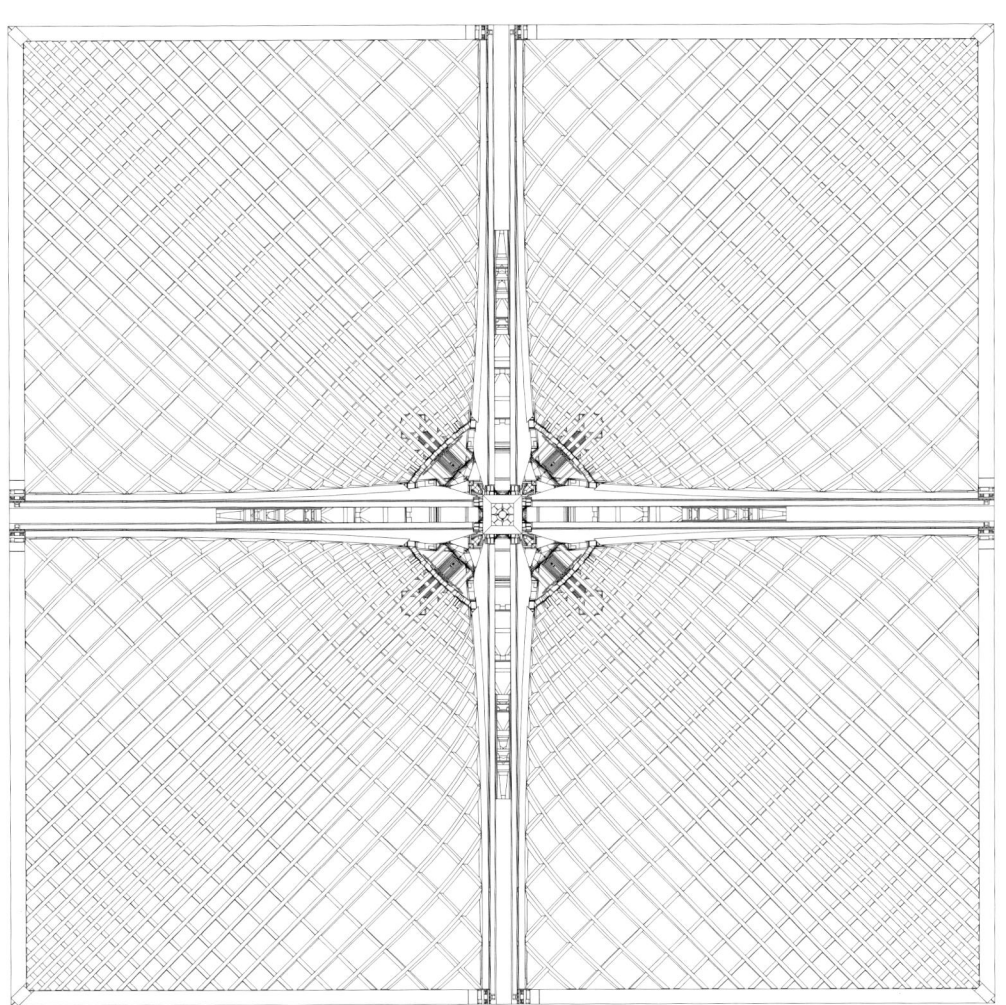

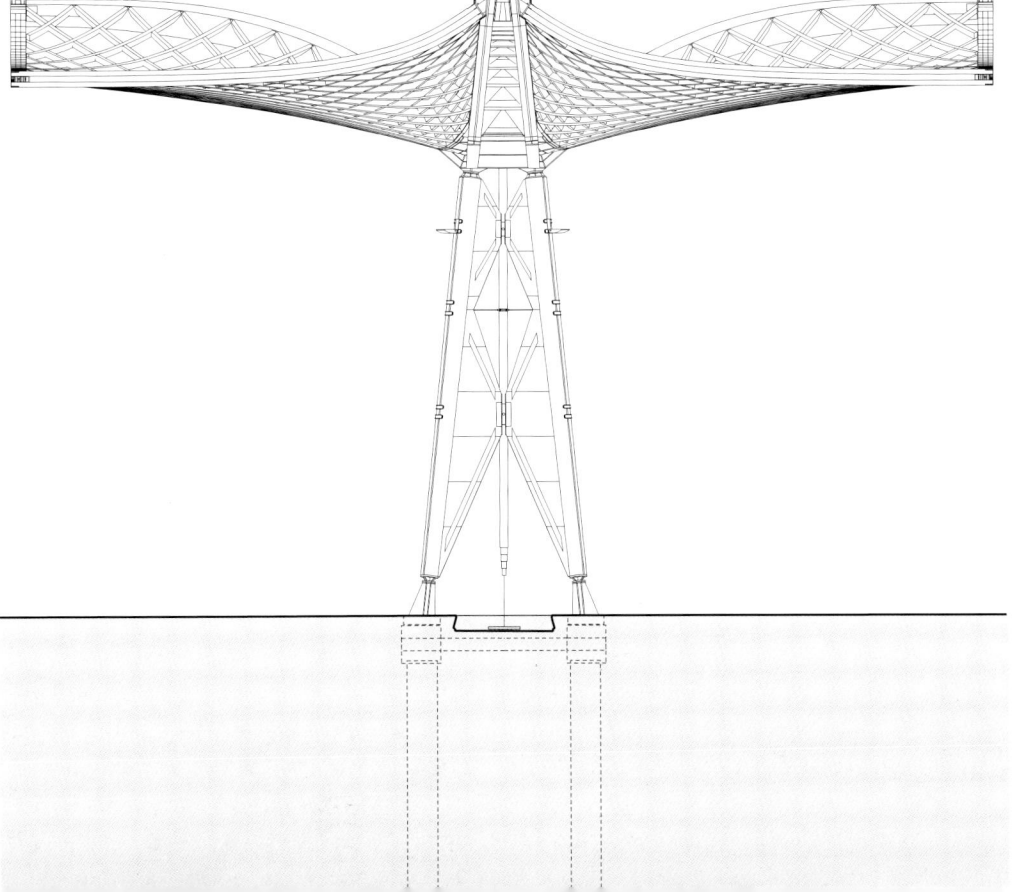

Die Farben der Hölzer

The Colours of the Timber

Rainer Wittenborn

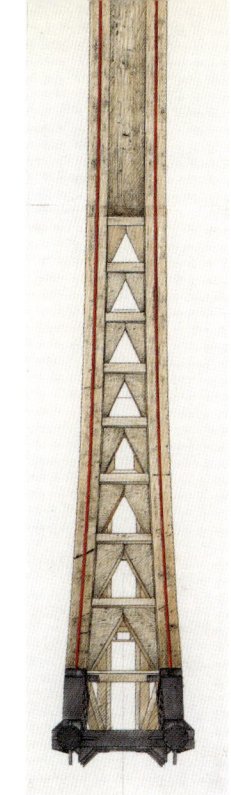

Je nach Material und Verarbeitung zeigen die verwendeten Hölzer ihre eigenen Töne und Farben. Die mächtigen Bäume – tragende Säulen des Daches – behalten die warme Grau-Skala, die nach Entrinden und Trocknen unter freiem Himmel entstanden ist. Geschnittene und gehobelte Teile wie Kragträger und deckende Blattschalen bis zu den eingespannten Holzverbundplatten zwischen den Stämmen finden im Zeitverlauf ihren adäquaten Ton im »hölzernen Farbfächer«.

Nur zwei Farben kontrastieren die Naturtöne des Materials Holz: das anthrazitfarbige Grau der verbindenden Stahlteile und die Farbe Rot – bei den eingeborenen Stämmen Amazoniens die Farbe des Lebens, in den Tempeln Südostasiens die Farbe der Säulen.

Ein kühles Rot legt den Kraftfluß in die Schnittfugen der Baumsäulen und läuft in den schwingenden Schattenfugen der Dachträger aus; ein warmes Rot faßt beidseitig die Aussteifungsrippen, welche die Linie der Konstruktion auf den Holzverbundflächen zwischen den Baumsäulen sichtbar werden lassen.

The timbers and wood products used in this project retain their natural colours and tones, which reflect the nature of the material itself and the way it was processed. The huge trees that form the supporting columns for the roof have their own warm-grey coloration – the result of a natural process of seasoning in the open air after the bark was stripped from the trunks. Cut and wrot elements such as the cantilevered trusses and the leaf-like shell canopies or the plywood sheets fitted between the stems of the towers will find their own appropriate colour in the "timber palette" in the course of time.

Only two colours are contrasted with the natural tones of the wood: the anthracite-grey of the steel connections, and a red tone that is regarded as the colour of life by native tribes in the Amazon basin, and which is also the colour of many columns in the temples of south-east Asia.

The cool red in the cut joints of the tree stems, which indicates the flow of forces in the columns, spreads out into the curving shadow joints of the roof beams. The bracing members to the tower are articulated on both sides by a warm red tone that visibly traces the structural lines on the plywood surfaces between the tree-columns.

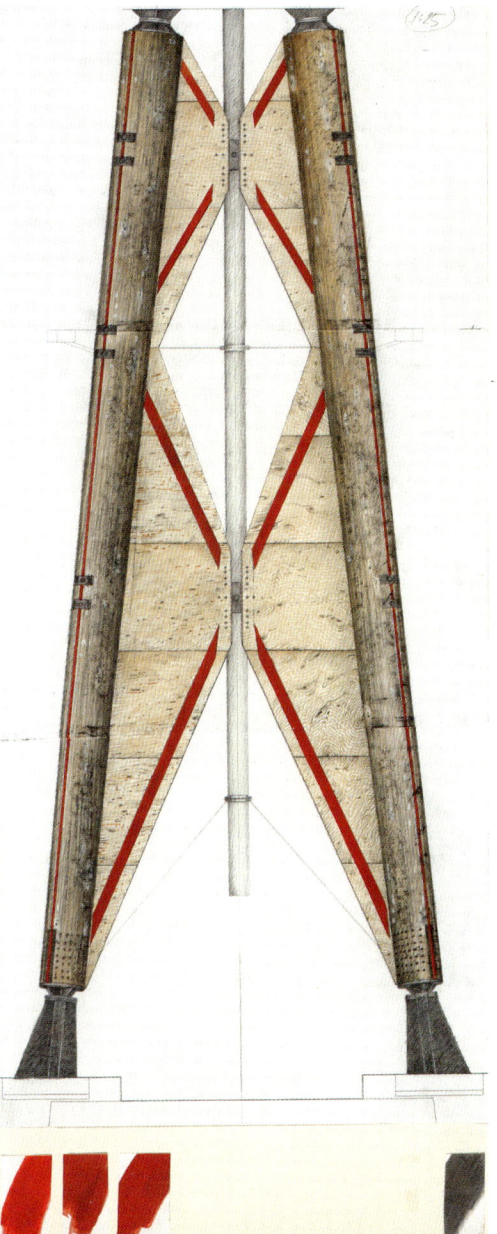

Zeichnung Rainer Wittenborn /
Drawing by Rainer Wittenborn

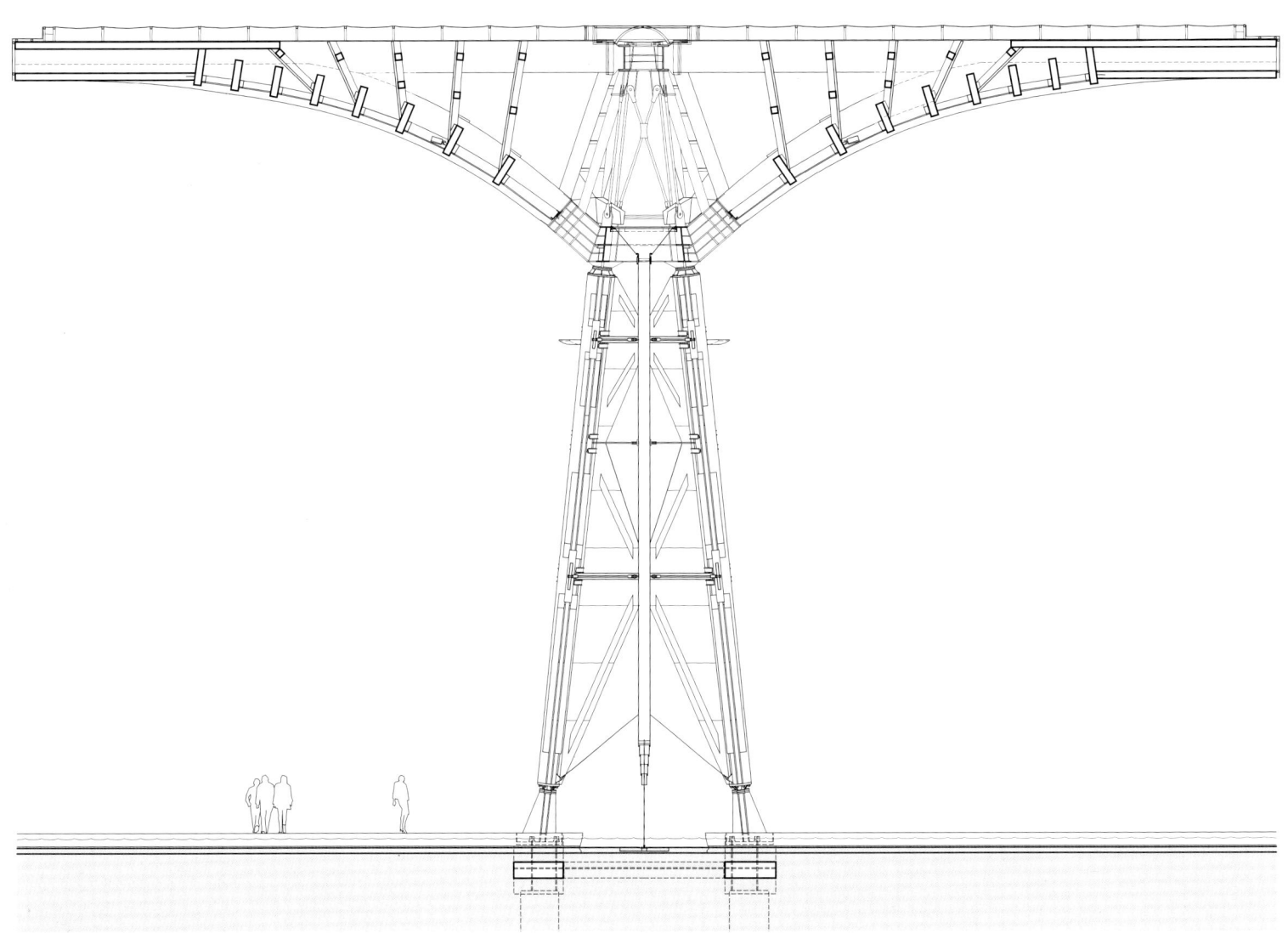

Querschnitt Maßstab 1:200 / Cross-section scale 1:200

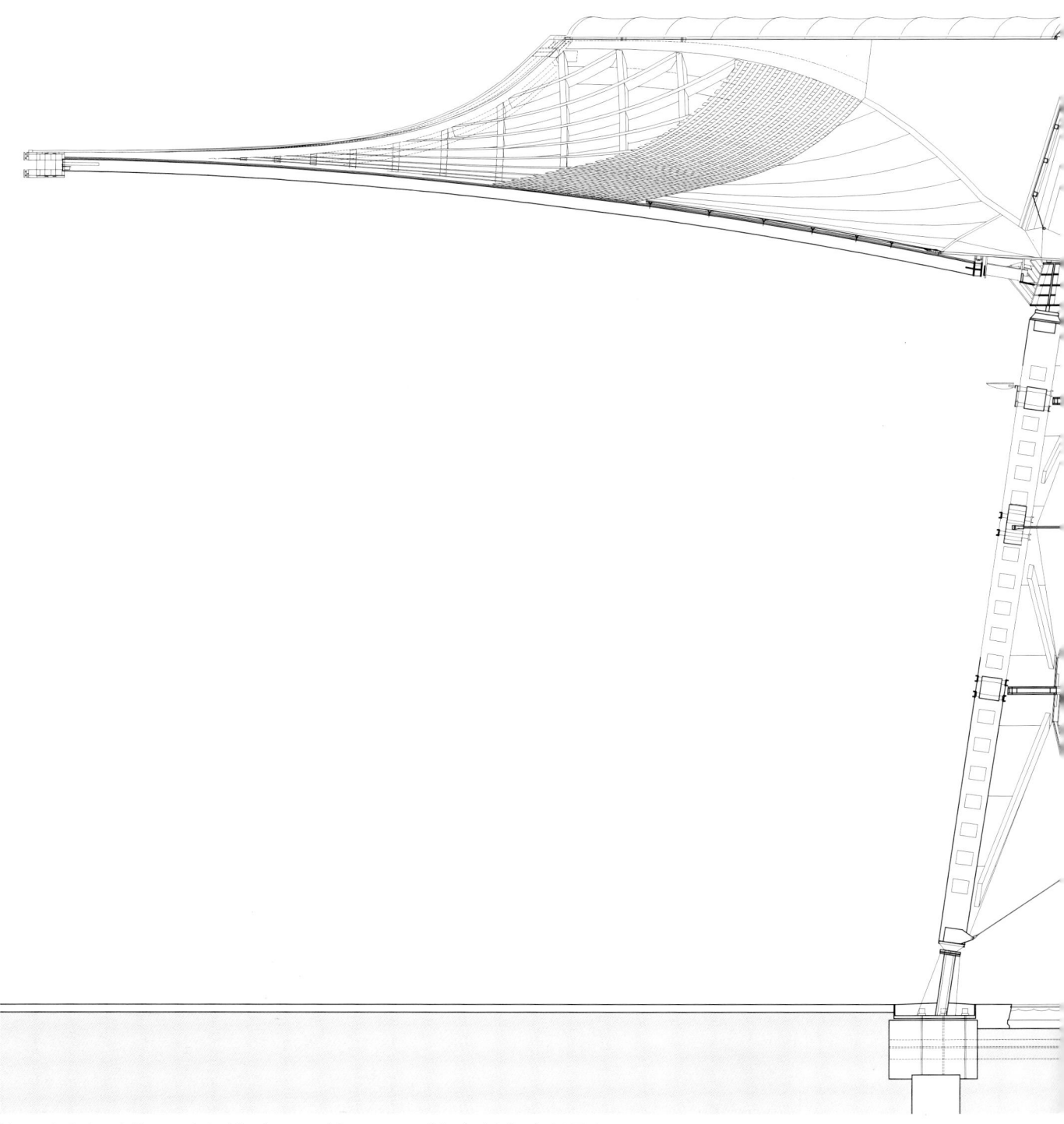

Diagonalschnitt mit Rippenschale, Membrane und Regenwassereinlauf Maßstab 1:150 /
Diagonal section, showing ribbed shell, membrane and rainwater drainage scale 1:150

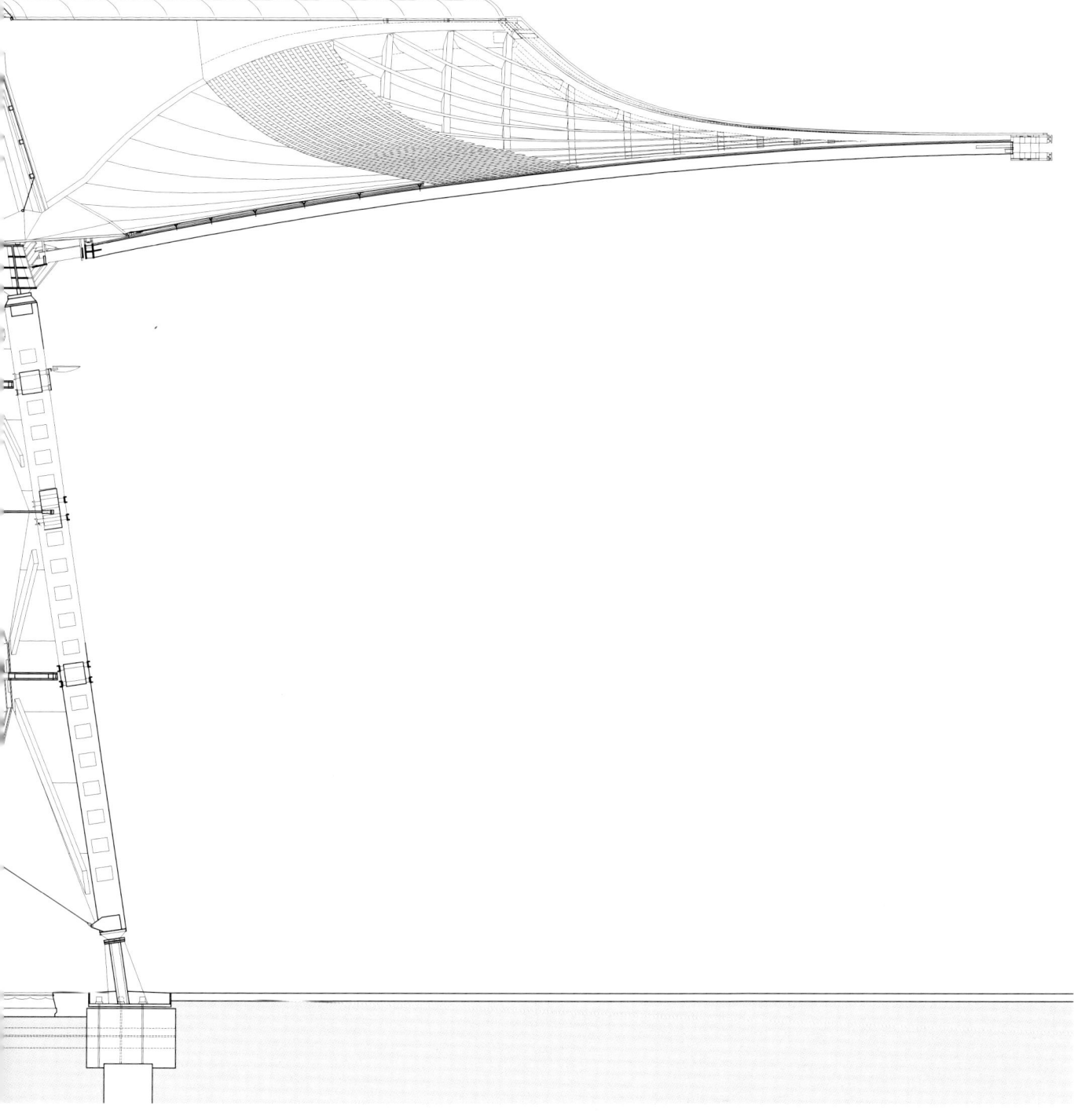

Die Membrane

The Membrane

Vorarbeiten zum Aufbringen der Membrane:
Setzen der Schaukelschrauben für die Seilkonstruktion

Preparatory work before laying membrane:
fixing eye screws for cable construction

Ausbreiten der Membrane / Laying out the membrane

Die Schalenflächen der Schirme sind mit einer transluzenten Dachhaut in Form einer Membrane überspannt, die nicht auf der Oberfläche der Schalung aufliegt. Die Membrane ist in der Fläche über Seile befestigt, die über den Längsrippen verlaufen und am Ende auf ihnen fixiert sind. Die Seile sind über Schraubösen direkt in die Längsrippen verankert. An den Rändern der Schalen ist die Membrane mit einer Klemmleiste gehalten.

Die Membrane wird mit einer Vorspannkraft von 1 kN/m aufgespannt. Sie wird in Führungstaschen an der Unterseite durch Seile gehalten; die Seile sind punktweise auf das Holzgitter verschraubt (Auflagerungsfunktion). Unter Belastung erhöht sich die Membranspannung, und es entstehen Seilkräfte. In der Fläche gleichen sich die Membranspannungen und Seilkräfte an den Verschraubungspunkten aus.

Am Rand werden die Membranspannungen und Seilkräfte direkt in die Längsrippen eingetragen und verursachen eine zusätzliche Normalspannung (Druck) sowie aufgrund der Exzentrizität ein zusätzliches Moment. Die Membrankräfte werden über eine Randbohle direkt in die einzelnen Bretter der Schalungslage eingetragen. Das Exzentrizitätsmoment wird über die Randbohle in den Untergurt der Kragträger bzw. in die Rippen am Randträger eingetragen.

Verwendetes Material:	Dach: PTFE/Glas
	Zwischenraum: ETFE
Einsatzgebiet:	permanente Bauten
Durchschnittliche Lebensdauer:	über 20 Jahre
Witterungsbeständigkeit:	sehr gut
Anschmutzverhalten:	sehr gut
Transluzenz:	sehr gut
Brandverhalten:	sehr gut

Anforderungen bezüglich Ökologie und Holzbau:
Recyclingfähigkeit / Humantoxizität / Umwelttoxizität / Energiehaushalt

ETFE-Folien bestehen aus Kohlenstoff, Wasserstoff und Fluor. ETFE-Folien werden ohne Additive hergestellt und verarbeitet, eine Steuerung der Farbgebung kann durch die Zugabe von Farbpigmenten erfolgen.

Langzeiterfahrungen für PTFE-Membrane und ETFE-Folien sind vorhanden. Für PTFE-Membrane ist eine Lebensdauer von über 20 Jahren gegeben; die Verarbeitung im Winter ist möglich; die Farbe ist zunächst hellbraun, bleicht aus und wird weiß.

ETFE-Folie wird zumeist als pneumatische zwei- oder mehrlagige Konstruktion eingesetzt. Die maximale Spannweite hängt von den äußeren Lasten ab und variiert zwischen 2 und 5 m. Die Länge der einzelnen Elemente betrug bisher bis zu 50 m.

Keine sichtbaren Abnutzungserscheinungen gibt es durch die Bewitterung. Die Haltbarkeit beträgt über 20 Jahre. Fluorpolymerfilm ist gegen alle Umwelteinflüsse besonders widerstandsfähig.

Aufgrund der ausgezeichneten antiadhäsiven Eigenschaften wird loser Schmutz durch den Regen weggespült. Die Wartungsintensität ist sehr gering. Inspektionen müssen höchstens im jährlichen Rhythmus durchgeführt werden.

ETFE-Folie ist wegen der geringen Dicke von etwa 0,2/0,9 mm, die eine sehr geringe Brandlast darstellt, als schwer entflammbar (B1/A2 nach DIN 4102) eingestuft.

Links: Montierte Membrankonstruktion
Rechts: Dachaufsicht Membrandach, Lage der Zuschnitte und Aufteilung der Lichtbänder (Ausschnitt)

Left: Detail of assembled membrane construction
Right: Top view of membrane roof, position of cuttings and distribution of rooflight strips (detail)

Ausbreiten der Membrane / Laying out the membrane

Regenwassereinlauf; Montage der Lichtbänder / Rainwater outlet; assembly of rooflight strips

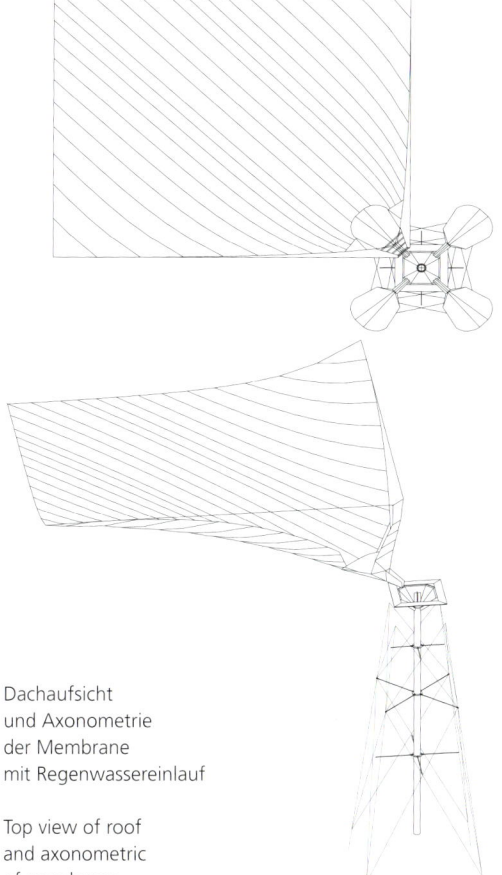
Dachaufsicht
und Axonometrie
der Membrane
mit Regenwassereinlauf

Top view of roof
and axonometric
of membrane
with rainwater drainage

Die Membrane ist zu 100% recyclierbar, bereits bewittertes Material jeden Alters kann durch Wiedereinschmelzen zu neuen Spritzgußteilen verarbeitet werden.

Die Extrusion der Folien benötigt ca. 1,1 kW/kg ET-Folie. Dies entspricht ca. 0,4 kW/m² bei einer Folienstärke von 0,2 mm.

Befestigung:
- Stahlseile auf den Längsrippen befestigt
- Ringschrauben (Ø 12 mm, Gewindelänge 65 mm) halten die Seile
- Membrane mit Taschen entlang der Längsrippen einkonfektioniert, in denen die Seile eingeführt sind.
- Aufbringen der Spannung an Kragträger und Randträger durch Spannen der Seile und der Membrane
- Schnelle Montage
- Montage im Freien möglich
- Keine Kollision mit bauseitigen Verbindungsmitteln
- Optimieren der Seilbefestigung durch Engersetzen der Ringschrauben in Bereichen mit hohen Lasten.

Die Holzkonstruktion kommt an keiner Stelle direkt mit der Membrane in Berührung und ist vor ständiger Durchfeuchtung durch anfallendes Tauwasser geschützt.

Technische Daten nach Angaben des Herstellers

The canopy shells are covered with a translucent skin or membrane which does not lie directly on the surface of the boarding. The membrane is fixed over its entire area by cables that run over the longitudinal ribs and are anchored to them by means of eye screws at the ends. At the edges of the shells, the membrane is fixed by clamping strips.

The membrane is prestressed with a tensioning force of 1 kN/m. It is held in position by cables threaded through linear pockets on the underside. The cables are fixed to the timber lattice by eye screws (bearing function). The tensioning of the membrane increases under loading, resulting in funicular forces. Over the entire area, the horizontal components of the membrane stresses and funicular forces balance each other out at the points of screw fixing. The membrane stresses and funicular forces are transmitted directly to the longitudinal ribs at the edges, where they result in additional normal stresses (compression) as well as an additional moment caused by their eccentricity. The membrane is fixed at the edges of the shells by planks that serve to transmit the loads from the membrane directly to the individual planks of the boarding layer. The eccentricity moment is transmitted via the edge planks to the lower chords of the cantilevered trusses and to the ribs along the edge beams.

Materials used: roof areas: glass-fibre/PTFE
intermediate roof strips: ETFE
Area of application: permanent buildings
Average life: more than 20 years
Resistance to weathering: very good
Resistance to soiling: very good
Translucency: very good
Fire resistance: very good

Aspects to be taken into account in terms of the environment and timber construction:
recyclability / human toxicity / environmental toxicity / energy balance

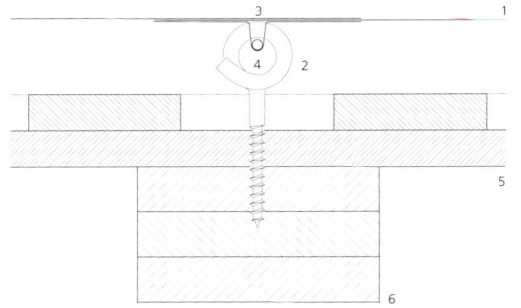

Detail Seilbefestigung der Membrane

1 Membrane
2 Holzbau-Ösenschraube
3 Schweißnaht
4 Befestigungsseil
5 Schalung, 2lagig
6 Längsrippe

Detail of membrane cable fixing

1 Membrane
2 Eye screw
3 Welded seam
4 Fixing cable
5 Two layers of boarding
6 Longitudinal rib

Montage. Rechts hinten Holzschale mit Transporttraverse

Assembly stage. In the background on the right: a timber shell with spreader for transport

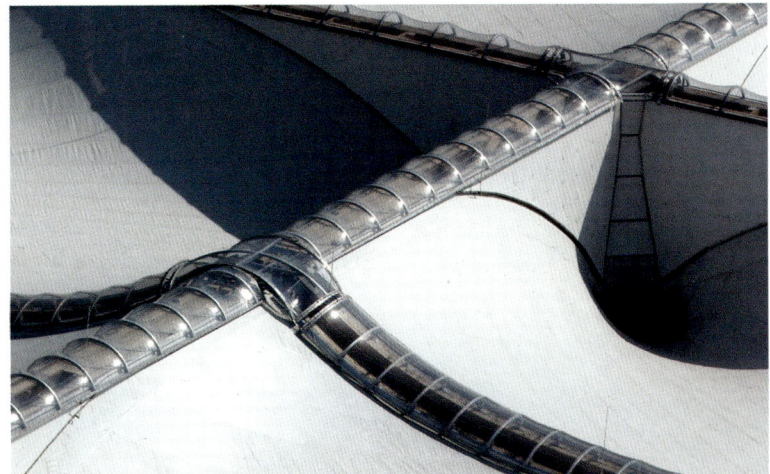

Tageslicht-Simulationen

Daylight Simulations

Thomas Kuckelkorn

ETFE synthetic sheeting consists of carbon, hydrogen and fluorine. It is manufactured and processed without additives. The coloration can be controlled by adding pigments.

Long-term experience in the use of PTFE and ETFE sheeting exists. The former has a guaranteed life of 20 years and can be laid in winter. The colour is initially pale brown, but the material bleaches out to become white. A longer life than that guaranteed is possible.

ETFE sheeting is normally used in pneumatic two- or multi-layer forms of construction. Maximum spans depend on the outer loading and can vary from 2 to 5 metres. The maximum length of the individual elements at present is up to 50 metres.

The material shows no visible signs of abrasion as a result of weathering. It has a life of more than 20 years. Fluoropolymer sheeting is particularly resistant to all environmental influences.

Because of the excellent non-adhesive properties of the sheeting, loose dirt is simply washed away by rainwater. The material has an extremely low maintenance intensity: inspections are necessary at most once a year.

In view of its thinness (0.2–0.9 mm), which represents an extremely low fire load, ETFE sheeting is graded as flameproof in accordance with German Standards (DIN 4102).

The membrane is 100 per cent recyclable. Material of any age, even where subject to heavy weathering, can be melted down again and processed into new injection-moulded elements.

The extrusion of ET sheeting requires roughly 1.1 kW/kg, or roughly 0.4 kW/m^2 of sheeting with a thickness of 200 μ (0.2 mm).

Fixing:
- steel cables are fixed on the longitudinal ribs
- the cables are secured by eye screws (12 mm dia. with 65 mm thread)
- membrane strip pockets are welded to the roof membrane along the longitudinal ribs; cables are threaded through the pockets
- cantilevered trusses and edge beams are stressed by tensioning the cables and membrane
- rapid assembly
- outdoor assembly possible
- no conflict with other connections on site
- optimization of cable fixings possible by closer spacing of eye screws in zones subject to greater loading

At no point is the timber structure in contact with the membrane, and it is protected against permanent exposure to condensation moisture.

Technical data provided by construction company

Um die natürlichen Belichtungsverhältnisse unterhalb der Außenüberdachung einschätzen zu können, wurden Simulationsrechnungen zum Bestimmen des Tageslicht-Quotienten durchgeführt. Für die Berechnungen verwendete man ein vereinfachtes Modell:

- die Dachhaut ist als plane Fläche angenommen;
- für das gesamte Dach wird ein einheitlicher Transmissionsgrad verwendet;
- durch den Dachaufbau bedingte Vorzugsrichtungen in der Transmission werden nicht berücksichtigt.

Die Größenordnung des zu erwartenden Transmissionsgrades der Dachfläche ist über den Anteil der opaken Flächen der Schirmkonstruktion abgeschätzt, wobei Korrekturfaktoren für den Höhenaufbau sowie für die Krümmung der Dachflächen angenommen sind.

In den dargestellten Berechnungsergebnissen variiert der Transmissionsgrad im Bereich zwischen opak (0% Transmission) und 10% Transmission.

Um den Einfluß weiterer raumbegrenzender Oberflächen auf die Tageslichtverhältnisse unterhalb der Überdachung zu beurteilen, wurden Berechnungen für verschiedene Reflexionswerte des Bodens und der Pavillonwandflächen angestellt.

Tageslicht-Simulationen
TD = Transmissionsgrad der Dachfläche
RD = Reflexionsgrad der Dachfläche
RB = Reflexionsgrad des Bodens
RP = Reflexionsgrad der Pavillons

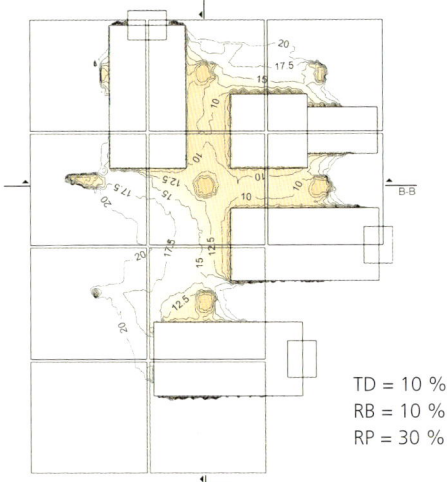

TD = 10 %
RB = 10 %
RP = 30 %

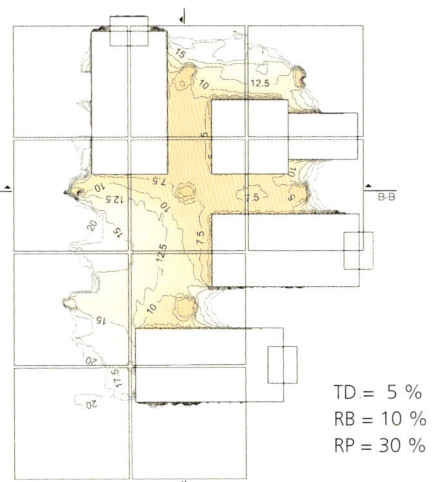

TD = 5 %
RB = 10 %
RP = 30 %

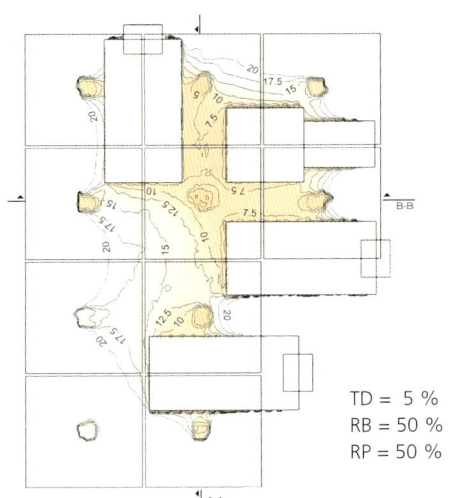

TD = 5 %
RB = 50 %
RP = 50 %

In assessing the natural lighting conditions beneath this open roof structure, simulation calculations were made to determine the daylight quotients. A simplified model was used for the calculation:

- the roof skin was assumed to be a planar surface;
- a uniform transmission rate was applied over the entire area of the roof;
- no account was taken of more favourable directions in the transmission of light, resulting from the form of the roof construction.

The degree of light transmission through the roof surface was estimated, taking into account the proportion of the canopy construction with an opaque surface and using assumed correction factors for the vertical dimension of the structure and the curvature of the roof.

The diagrams show the resulting daylight distribution for transmission factors of the roof skin ranging from 0 per cent (opaque areas) to 10 per cent.

To assess the influence of other space-enclosing surfaces on the daylight conditions beneath the roof, calculations were made for various reflection values of the ground and the pavilion walls.

Daylight simulations
TD = light-transmittance values of roof
RD = light-reflection values of roof
RB = light-reflection values of paving
RP = light-reflection values of pavilions

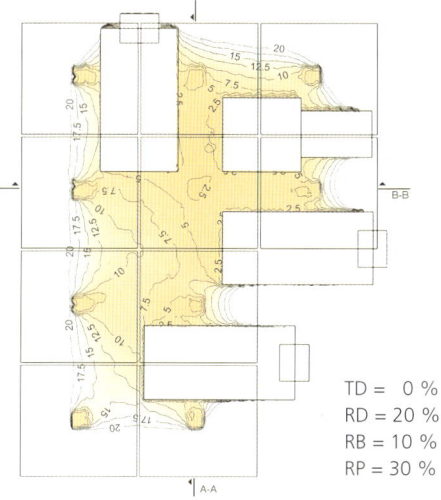

TD = 0 %
RD = 20 %
RB = 10 %
RP = 30 %

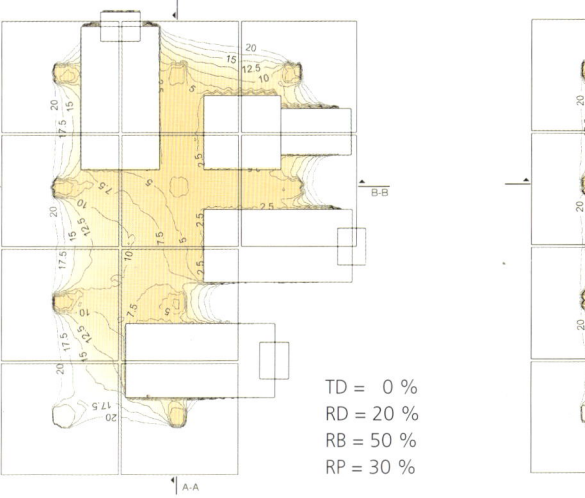

TD = 0 %
RD = 20 %
RB = 50 %
RP = 30 %

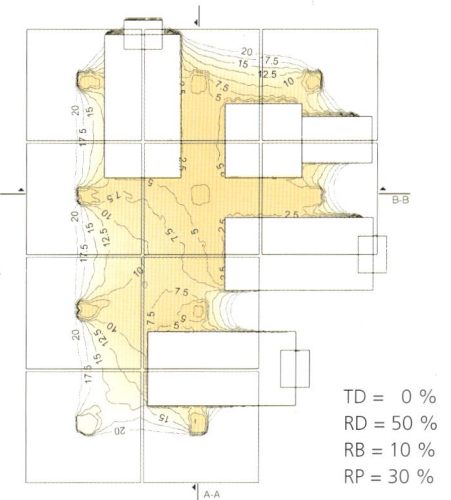

TD = 0 %
RD = 50 %
RB = 10 %
RP = 30 %

Weißtannen aus dem Schwarzwald

Silver Firs from the Black Forest

Christoph Hoffmann, Gerhard Rieger, Andreas Schabel

Weißtanne im Februar 1999

Silver fir in February 1999

Messen mit Ultraschall

Ultrasonic measurement

Die Weißtanne (abies alba) ist eine Charakterbaumart der europäischen Bergmischwälder. Aus der artenreichen Gattung der Tannen ist sie der einzige Vertreter in Mitteleuropa. Von den anderen bei uns vorkommenden Nadelbaumarten mit nach unten hängenden und als ganzes abfallenden Zapfen ist sie an den aufrecht auf den Zweigen stehenden Zapfen zu unterscheiden, die nach der Reife schuppenweise abfallen. Man wird deshalb nie einen echten Tannenzapfen auf dem Waldboden finden.

Baden-Württemberg ist das tannenreichste deutsche Bundesland. Der im Land stehende Vorrat an Tannenholz beträgt derzeit rund 47 Millionen Kubikmeter. Bemerkenswert ist die Verteilung nach Stärkestufen. Knapp ein Drittel des Holzvorrates ist in Bäumen konzentriert, die in 1,3 m Höhe einen Durchmesser von mehr als 50 cm aufweisen. Das sind 10,8 Millionen Kubikmeter Tannen-Starkholz. Damit ist die besondere Eignung der Tanne zur Starkholzproduktion belegt. Geliefert wurden die Bäume für das EXPO-Dach aus dem Kommunal- und Privatwald.

Die besonderen ökologischen Eigenschaften der Weißtanne sind:
- Hohe Stabilität gegenüber Sturmwurf durch ihr tiefgreifendes intensives Wurzelwerk,
- wesentlich höhere Schneebruchresistenz als die Fichte,
- keine alters- oder standortbedingte Stammfäule,
- gutes Ausheilen von Verletzungen,
- stabilisierende Wirkung auf den Kreislauf der Nährelemente, besonders an basenarmen Standorten,
- aufgrund ihrer Schattentoleranz besondere Eignung zum Schaffen strukturreicher Bestände.

Damit ist die Tanne die ideale Baumart für alle Formen der Dauerwaldwirtschaft (»Plenterwirtschaft« und »Femelwirtschaft«), mit denen Kahlschläge vermieden werden.

Holz, Klima, Kohlendioxid

Die Neubauten auf dem EXPO-Gelände orientieren sich – und dies ist ein wesentliches Novum im Vergleich zu allen bisherigen Weltausstellungen – am Kriterium der Nachhaltigkeit. So liegt es nahe, den Baustoff Holz als nachwachsenden Rohstoff für die Konstruktion dieses Großdachs einzusetzen.

Die Fähigkeit von Bäumen, das Gas CO_2 aus der Luft zu binden, ist für die Entwicklung des Klimas und damit der Umwelt zukünftig von vermehrter Bedeutung. Will man den CO_2-Gehalt der Luft nicht weiter ansteigen lassen, empfiehlt sich die Verwendung von Biomasse als Ersatz für fossile Brennstoffe, die letztlich Verursacher des Treibhauseffekts sind. Noch wichtiger ist jedoch die Holzverwendung als vielfältiger Rohstoff. Denn solange Holz nicht verbrannt wird oder verrottet, ist das CO_2 der Luft entzogen und festgelegt. So stellt jeder Balken eines Hauses, der sich über Jahrhunderte erhalten kann, eine CO_2-Festlegung dar. Je mehr Holz in den Wäldern produziert, anschließend genutzt und im Hochbau verwendet wird, desto besser ist dies für unser Klima. Auch die Waldnutzung und pflegliche Forstwirtschaft ist für die Klimaverbesserung von Bedeutung. Werden Bäume zu alt, von Insekten und Pilzen befallen, vermodern sie vor Ort und das dort in Bäumen festgelegte CO_2 wird wieder freigesetzt. Gesunde, produktive Wälder und die vermehrte Verwendung von Holz als Rohstoff bringen Entlastung für das Weltklima.

Holz als Baustoff für Tragkonstruktionen

Die Qualität von Holz für tragende Zwecke wird traditionell visuell beurteilt. Kriterien sind die Ästigkeit (Größe und Häufigkeit von Ästen, bezogen auf die Querschnittsabmessungen), die Breite der Jahresringe und die Faserabweichung (Abweichung der Richtung der Holzfasern von der Längsachse des zu klassifizierenden Querschnitts) sowie Schädigungen infolge von Rißbildung, Insekten und Pilzen. Die Klassifizierung hängt jedoch auch von der subjektiven Beurteilung des Sortierers ab.

Vorauswahl der Bäume im Wald

Alle 40 Stämme mußten sämtliche Anforderungen an die Holzqualität erfüllen. Als Kriterien hierfür sind Äste, Krümmungen, Jahresringbreite und Mindestrohdichte zu beurteilen.

Maschinelle Verfahren zum Sortieren von Rundholz oder Schnittholz mit größeren Querschnitten existieren nicht. Für die Auswahl und Qualitätsbestimmung des Rundholzes für die Stützen der Turmkonstruktion des EXPO-Daches kam daher nur visuelle Begutachtung in Frage.

Zum Herstellen der Stützen in der Turmkonstruktion benötigte man Stämme mit mindestens 72 cm Durchmesser am Zopf (oberes, dünneres Ende eines Stammabschnittes) und 90 cm Durchmesser am Stock (unteres, dickeres Ende eines Stammabschnittes). Unter Berücksichtigung einer ausreichenden Überlänge für den Abbund, d. h. das Herstellen der Bauteile mit allen Fräsungen und Bohrungen, mußten die Stammabschnitte eine Länge von mindestens 18 m aufweisen.

Sowohl im Süd- als auch im Nordschwarzwald konnten Stämme entsprechender Dimension gefunden werden. Man entschied sich für den Südschwarzwald, da in den Forstbezirken Schopfheim und Todtmoos wohl die höchste Konzentration dieser Waldgiganten in Westeuropa vorhanden ist.

Diese Stämme fanden in früheren Jahren wegen ihrer Wetterfestigkeit, ihres geraden Wuchses und ihrer Vollholzigkeit häufig als Masten im Schiffsbau für Großsegler Verwendung. Sie weisen ferner gleichmäßige Jahresringe auf und neigen nur wenig zu Druckholzbildung[1] und Drehwuchs[2].

Bei Tannen größeren Durchmessers treten bisweilen Naßkerne mit hohen Holzfeuchten auf (100% bis 200%). Üblicherweise ist Kernholz bei frisch gefällten Bäumen weniger feucht (30% bis 50%) als das feuchte Splintholz der Randzonen, das durchaus Holzfeuchten bis 200% aufweisen kann.

---

[1] Bei einseitiger Belastung, z.B. durch Neigung der Stammachse oder durch häufige Windbeanspruchung aus einer Richtung, bilden Bäume sog. Reaktionsholz mit verändertem Zellaufbau und damit veränderten Eigenschaften. Bei Nadelholz bildet sich auf der druckbeanspruchten Seite das Druckholz, bei Laubbäumen auf der zugbeanspruchten Seite das Zugholz.

[2] Bei Drehwuchs verlaufen die Holzfasern spiralförmig und weisen daher einen Winkel zur Längsachse des Stammes auf. Drehwuchs führt bei Schnittholz zu Abweichungen der Faserrichtung von der Bauteilachse, was ungünstige Auswirkungen auf die Festigkeit hat.

Fällen des Baumes

Felling a tree

Messen mit Ultraschall

- Die Holzgüte wurde am stehenden Stamm mittels Ultraschall durch das Institut für Holzkonstruktionen der Ecole Polytechnique Fédérale de Lausanne (EPFL) gemessen (optische Messung des Stammdurchmessers in 18 m Höhe mit Theodolit und Beurteilen starker Krümmungen und Astigkeit).
- Wie ist die Holzfeuchte der Tanne zu beurteilen und welche Auswirkungen kann diese haben (Naßkern)? Kann durch ein Längs-Auftrennen der Stämme ein günstiges Verhalten (Statik, Optik, Technik) erreicht werden? Aufgrund der vorliegenden forstlichen Erfahrungen wurde die Möglichkeit ausgeschlossen, daß das Holz wegen seiner hohen Ausgangsfeuchte faulen oder unangenehm riechen würde, was von Skeptikern zunächst behauptet wurde.

Die Stammdurchmesser bestimmte man in Brusthöhe, ca. 1,30 m über dem Boden, direkt durch Kluppen (Schiebelehren zum Messen von Baumdurchmessern). Eines der wesentlichsten Auswahlkriterien war der Zopfdurchmesser, der natürlich nicht direkt zu messen war. Mitarbeiter der Forstlichen Versuchsanstalt in Freiburg nahmen deshalb Einschätzungen der Durchmesser in 18 m Höhe mit Hilfe wissenschaftlicher Erfahrungswerte über den Durchmesser in Brusthöhe und zusätzlich vermessungstechnisch mit einem Theodolit vor.

Die nach dem allgemeinen Erscheinungsbild und den Abmessungen in Frage kommenden Stämme wurden markiert und anschließend mit Ultraschall untersucht. Mit dieser im deutschsprachigen Raum noch kaum eingesetzten Methode verhindert man zum einen, daß Bäume mit Kernfäule eingeschlagen werden (durch eine längere Wegstrecke der Schallwellen um die Faulstellen herum erhöht sich die Schall-Laufzeit durch den Querschnitt). Zum anderen bestimmt man von den vorausgewählten Bäumen die besseren. Die Beurteilung geschah durch Vergleich der gemessenen Ultraschall-Geschwindigkeiten. Die Geschwindigkeit ist bei höherer Rohdichte größer.

An jedem vorausgewählten Baum wurden Ultraschallmessungen in radialer Richtung durch den Stammfuß gemacht. Das Ergebnis bestätigte das Verfahren: Unter den gefällten Bäumen befanden sich keine Exemplare mit Kernfäule.

Alle Stämme wurden sektionsweise auf Drehwuchs vermessen. Zusätzlich wurden Messungen der Halbmesser und jeweils vier Laufzeit-Messungen in Stammlängsrichtung an den gefällten Bäumen durchgeführ

Einschlagen, Rücken und Lagern der Stämme

Das Trocknungsverhalten von Rundhölzern dieser Dimensionen ist weitgehend unbekannt. Um vermehrte Rißbildung zu vermeiden, wurde eine ausschließlich natürliche Trocknung gewählt. Den Eigenschaften ›Druckholz‹ und ›Drehwuchs‹ kam infolge eines veränderten Schwindverhaltens bei der späteren Auswahl der Stämme für die Konstruktion besondere Bedeutung zu.

Um die Zeit bis zum Einbau für natürliches Trocknen möglichst gut auszunutzen, wählte man die Bäume noch während des Winters bei einer extremen Schneelage im Jahr 1999 aus.

Die Bäume wurden vor Beginn der Wachstumsperiode geschlagen und waren so weniger anfällig für Schädigungen durch Pilze. Die Rückearbeiten, das heißt der Transport der Bäume aus dem Wald, und der Transport zum Lagerplatz mußte vor dem möglichen Auftreten holzschädigender Insekten geschehen.

Die Stämme wogen pro Stück zwischen 9 und 15 Tonnen. Der größte Hiebsdurchmesser lag bei 1,45 Meter; der höchste Baum maß 51 Meter. Das Fällen und der Transport der Stämme aus dem Wald war aufgrund der Dimensionen und des hohen Gewichtes ungewöhnlich und schwierig. Zum Rücken im Wald waren bis zu drei Forstschlepper nötig. Je Lkw konnte nur ein Stamm aus dem Wald abgefahren werden. In Gersbach war ein etwa 350 x 50 Meter großer Lagerplatz eingerichtet. Die Tannen wurden dort mit 400 bar Wasserdruck entrindet und mit einer Motorsäge (1,80 Schwert und zwei Antriebe) längs in zwei Hälften aufgesägt.

Vorbereitende Arbeiten am Lagerplatz

Das Entrinden der Stammabschnitte folgte unmittelbar nach Anliefern der Stämme am Lagerplatz. Die durch die Art der Entrindung mit Hochdruck-Wasserstrahl erzielte glatte und gleichmäßige Oberfläche fand Gefallen, so daß man auf das zunächst vorgesehene Abfräsen verzichten konnte.

Die weitere Bearbeitung der Stammabschnitte wurde unter Beachtung gestalterischer, statischer, konstruktiver, montage- und ausführungstechnischer Gesichtspunkte diskutiert. Hier ergab sich auch die bereits erwähnte Längs-Halbierung der Stämme. Die Schnittführung orientierte sich dabei an einer mit Hilfe einer Schlagschnur angebrachten Markierung.

Um das Austrocknen zu unterstützen, wurden die Stammhälften mit dazwischen liegenden Hölzern gelagert, durch die ein Durchlüften gewährleistet wird, und mit Spanngurten wieder zusammengespannt. Um den gleichen Trocknungsgrad beider Stammhälften zu erreichen, wendete man die Stämme nach der halben Lagerungszeit.

Auf dem Lagerplatz wurden sämtliche Stämme nochmals durch die EPFL mit Ultraschall mehrmals in Längsrichtung gemessen.

Das Einstufen in Sortierklassen erfolgte anhand der Kenngrößen: Ästigkeit, Jahresringbreite und Drehwuchs durch Prof. Dr.-Ing. Kessel, TU Braunschweig (siehe Seite 38f). Von jedem Stamm wurden Stammscheiben zum Bestimmen der Rohdichte und der Jahrringbreiten abgeschnitten und untersucht. Ergänzend dazu wurden 78 Prüfkörper für Bruchversuche hergestellt. Zum Überprüfen der Berechnungsannahmen wurden daran die Biegefestigkeit und das Elastizitätsmodul bestimmt.

Die deutsche Forst- und Holzwirtschaft stellt mit der Realisierung des EXPO-Dachs ihre Kompetenz sowohl mit den eingesetzten Materialien als auch durch die Bauleistung unter Beweis. Eine breite Gruppe von Fachleuten und Mitarbeitern ist daran beteiligt, beginnend beim Forst, über die Sägewerke und das Anfertigen von Brettschichtholz, über den Handel bis zu den Holzbauunternehmen.

The silver fir (abies alba) is a species typical of the mixed mountain forests of central and southern Europe. Like all species of *abies*, the silver fir is distinguished by its cones, which grow upright on the branches and shed their individual scales when they are ripe. In contrast, the cones of other coniferous trees native to this part of Europe hang from the branches and fall to the ground whole. That is the reason why one will never find a ripe fir cone on the floor of a forest.

Baden-Württemberg possesses the greatest stock of standing silver fir trees of any state in Germany. The present volume of fir timber growing in that state is about 47 million cubic metres. A further striking statistic is provided by a grading according to stem thickness. Nearly a third of the timber reserves consist of trees with a trunk diameter of more than 50 cm at a height of 1.30 m above the ground. That represents 10.8 million cubic metres of large-sized silver fir timber, which demonstrates the special suitability of this type of tree for heavy timber production. The trees for the EXPO roof came from publicly and privately owned forests.

The special ecological characteristics of the silver fir are:
- great stability and resistance to uprooting by gales as a result of its deep, dense root system;
- considerably higher resistance than the common European spruce to damage caused by snow loads;
- no trunk rot resulting from age or locational factors;
- good wound-healing capacity;
- the stabilizing effect it has on the cycle of nutritive elements, especially in low-alkali locations;
- its suitability to the creation of mixed uneven stands as a result of its tolerance of shade.

The fir is, therefore, an ideal species for all forms of long-term forestry (selective felling and thinning-out systems), where one wishes to avoid clear-cutting whole areas.

Wood, climate, carbon dioxide

The new buildings on the EXPO site are based on a principle of environmental sustainability – a major innovation in comparison with all previous world exhibitions. It seemed only logical, therefore, to use timber, as a regenerable raw material, for this large roof structure.

The ability of trees to absorb CO_2 from the air is of special significance in terms of future climatic developments and of the environment in general. If the content of CO_2 in the air is not to increase further, it makes sense to use biomass as a replacement for fossil fuels, which are ultimately the cause of the greenhouse effect. Even more important is the use of wood in its many different forms as a raw material. As long as wood is not burned or does not decay, the CO_2 it contains is removed from the air and bound up in the material. Every beam in a house that has survived the centuries represents a certain amount of CO_2 that has been absorbed. The more timber we produce in our forests and subsequently use in building construction, the better it is for the climate. The cultivation and good management of forests are also of great significance in improving climatic conditions. If trees grow too old, however, or become infested with insects or fungi, they may rot where they stand, and the CO_2 they contain will be released into the air again. Healthy, productive forests and the increased use of wood as a raw material can, therefore, help to improve the global climate.

Timber as a material for load-bearing structures

Traditionally, the quality of timber used for load-bearing structures is assessed visually. The criteria include knot ratio (the size and frequency of knots in relation to the cross-sectional dimensions), the width of the annual rings, deviations in fibre orientation (deviation of the direction of wood fibres from the longitudinal axis of the cross-section), and damage to the stem caused by cracking, insects and fungi. The classification will also depend on the subjective judgement of the person grading the timber.

There are no machine grading systems for round timber stems or cut timber of larger cross-sections. In determining the quality of the tree stems and selecting them for the columns of the EXPO roof towers, therefore, only a visual evaluation was possible.

For the tower column construction, tree stems were required with a diameter of at least 72 cm at the head of the cut section of the trunk and 90 cm at the stump or base. Allowing sufficient additional length for cutting and waste (i.e. the manufacture of the structural elements after all cutting, shaping and boring), the cut stems had to have a length of at least 18 m.

Trees with the requisite dimensions can be found in the northern and southern Black Forest. A decision was made in favour of the southern Black Forest, since probably the highest concentration of these giant trees in western Europe is to be found there in the forestry districts of Schopfheim and Todtmoos.

Because of their weatherproof qualities, their straight growth and their density, trees of this kind were frequently used in earlier times for the masts of ships. They also have an even annual ring growth and a minimum tendency to the formation of compression wood[1] and twisted grain[2].

Silver fir trees of larger diameters sometimes have a wet core with a high relative moisture content in the heartwood (100–200 per cent). As a rule, the heartwood of freshly felled trees has a lower moisture content (30–50 per cent) than the moist sapwood in the outer layers, which may have a relative moisture content of up to 200 per cent.

Preselection of trees in the forest

All 40 tree stems required for the roof towers had to meet the entire list of requirements in terms of timber quality. The criteria for the selection included knots, curvature and twisting, annual-ring width and minimum relative density.

1 Where one-sided loading occurs (e.g. as a result of the trunk axis being inclined or of frequent wind loading from one direction), trees form so-called "compression wood", which has a different cell structure and thus different properties. In the case of wood from coniferous trees, compression wood develops on the side subject to compression; while with deciduous trees, tension wood develops on the side subject to tension.
2 Where twisted grain occurs, the wood fibres have a spiralling form and thus lie at an angle to the longitudinal axis of the trunk. In cut timbers, twisted grain results in deviations in the direction of the fibres from the main axis of the constructional component, which has a negative effect on its strength.

Entrinden mit 400 bar Wasserdruck /
Removing bark with 400-bars water pressure

Halbieren der Stämme /
Cutting trunks in half

Präzise Schnittführung /
Precision sawing

◁◁ Transport der Baumgiganten zum Lagerplatz /
◁◁ Transport of giant tree trunks to storage site

◁ Lagerplatz bei Gersbach/Schwarzwald; im Vordergrund geschälte Stämme /
◁ Storage site near Gersbach in Black Forest; in the foreground: tree trunks with bark removed

Ultrasonic measurement

- Measurements were made by the Institute for Timber Structures of the Ecole Polytechnique Fédérale de Lausanne (EPFL) to determine the timber quality of the trees while still standing, using ultrasonic equipment (plus visual measurement of the diameter of the trunk at a height of 18 m above ground level and assessment of heavy twists and knot sizes).
- How can the moisture content of the fir trees be judged, and what effects can this have (e.g. wet core)? Could the behaviour of the stems be positively influenced (structurally, visually and technically) by cutting the trunks in half lengthwise? Existing forestry experience ruled out the possibility of the timber rotting or giving off an unpleasant smell as a result of its high initial moisture content – as was at first asserted by sceptics.

The diameter of the trees at chest height (roughly 1.30 m above the ground) was determined by direct means, using adjustable calipers specially made for this purpose. One of the main criteria for selection was the diameter at the head, which it was not, of course, possible to measure directly. Assistants of the Forestry Research Institute in Freiburg made estimates of the diameter at a height of 18 m based on scientific empirical values in relation to the diameter at chest height. These calculations were supported by additional measurements with a theodolite.

The trees considered most suitable – based on an assessment of their overall appearance and dimensions – were marked and subsequently investigated with ultrasonic equipment. This technique, which has scarcely been used before in German-speaking countries, helps to avoid felling trees with heart rot. (Since the sound waves have to travel a greater distance around the rotten sections, the time they take to pass through the cross-section is longer.) The method also enabled the best specimens to be chosen from among the preselected trees. This evaluation was based on a comparison of ultrasonic speeds (the greater the relative density, the faster the soundwaves travel).

Ultrasonic measurements were made in a radial direction through the base of every tree to be felled. The results confirmed the efficiency of this process: among the trees actually felled, there was not one with heart rot.

All trees were also measured at regular intervals along their length for twisted grain. In addition, measurements of the half-stems were taken, and the duration of the ultrasonic signal was measured at four points along the longitudinal axes of each of the trunks of the felled trees. From the stems that had been cut and were deemed suitable, those with the better measurement data were finally selected.

Felling, transporting and storing the tree stems

Little is known about the behaviour of round stems of such dimensions when they are seasoned. To avoid excessive cracking, an entirely natural seasoning process was chosen. In view of the different shrinkage behaviour of compression wood and wood with twisted grain, the identification of these phenomena was an important aspect in the subsequent selection of the stems for the roof construction.

In order to make the best use of the time remaining before the erection of the roof to dry the timber, the trees were selected during the winter of 1999 when deep snow lay on the ground.

The trees were felled before their growth period commenced; they were, therefore, less vulnerable to fungus attack. The removal of the logs from the forest and their transport to the storage area had to take place before destructive insects could damage the timber.

Each of the trunks weighed between 9 and 15 tonnes. The greatest cut-off diameter was about 1.45 m, and the tallest tree 51 m high. The felling and skidding of the trunks from the forest was exceptionally difficult because of the dimensions and the great weight. Up to three skidding engines were needed to drag the stems through the forest, and the lorries were able to transport only one trunk at a time. A roughly 350 x 50 m storage area was set up in Gersbach in the Black Forest, where the bark was stripped from the fir trees by means of high-pressure water jets (400 bars) and the stems were cut in half lengthwise by a power saw with a 1.80 m blade driven by two motors.

Preparatory work at the storage site

The process of stripping the bark from the logs took place immediately after their arrival at the storage site. The smooth, even surface achieved by the method of removing the bark was very pleasing, so that it was not necessary to grind the surface as originally foreseen.

Further treatment of the stems was discussed in the context of the design – for structural and constructional purposes, and in terms of the assembly and technical execution. The cutting was performed with the help of chalk-line markings.

To support the seasoning process, wood battens were inserted between the halves of the stems, which were then strapped together. This allowed air to circulate around and between the sections. Halfway through the seasoning period, the stems were turned to ensure an even drying of both halves.

On the storage site, all stems were subjected to further ultrasonic tests over their entire length. The measurements were again carried out by the EPFL.

The logs were later graded into different classes by Prof. Dr.-Ing. Kessel of the University of Technology, Brunswick, on the basis of nominal sizes, knot ratio, annual-ring widths and spiral grain (see page 38f.). Cross-sectional slices were cut from every stem and investigated to determine the exact density and the annual-ring width. In addition to all this, 78 specimens were prepared for testing the material to failure point. The bending strength and Young's modulus were determined in order to test the assumptions made in the calculations.

With the realization of the EXPO roof, German forestry and the timber industry have demonstrated their competence, in terms of the materials used and the constructional achievement this structure represents. A large number of experts and organizations were involved, representing the forestry sector, the sawmills and plants where laminated timber is produced, the timber trade and the timber construction firms.

Laborversuche: Festigkeit und Steifigkeit der Tannenstämme

Laboratory Tests: Strength and Rigidity of the Silver Fir Stems

Martin H. Kessel

Getrocknete, an der Oberfläche ergraute Stammhälften

Seasoned half-stems with greying surface

Zum Nachweis der Tragfähigkeit der aus vier Tannenstämmen bestehenden Mastkonstruktion eines Schirmes konnte auf die in den Baunormen für Nadelholz angegebenen Materialkennwerte nicht ausreichend zurückgegriffen werden, weshalb zusätzliche Untersuchungen erforderlich waren; die Abmessungen der verwendeten Tannenstämme weichen wesentlich von den Abmessungen von üblicherweise zum Bau verwendetem Nadelholz ab. Zunächst mußten für alle Tannenstämme Holzdichte, Ästigkeit und Faserneigung in bezug auf die Stammachse bestimmt werden. Die Meßergebnisse wurden dann daraufhin überprüft, ob ihre Größen mit den entsprechenden bekannten Werten für übliches Nadelholz vergleichbar sind und eine Einteilung der Tannenstämme in die für Nadelschnittholz definierten Sortierklassen erlaubten. Da die Einteilung in die Sortierklassen darüber entscheidet, welche Tragfähigkeit die Tannenstämme besitzen, wurde anschließend stichprobenartig Festigkeit und Steifigkeit in Bruchversuchen bestimmt, um den Zusammenhang von Sortierklasse und Biegefestigkeit und Biegesteifigkeit der Stichprobe herzustellen.

Zum Bestimmen der Holzdichte wurden von 51 Tannenstämmen 10 bis 15 cm dicke Stammscheiben (Bild 1) vom oberen Ende (Zopfende) der auf etwa 17 Meter abgelängten Stämme abgeschnitten. Die Stammscheiben hatten einen Durchmesser von 66 bis 86 cm. Die Verteilung der Anzahl der Jahrringe ist ebenfalls in Bild 1 dargestellt. Aus der Jahrringanzahl läßt sich das Alter des Baumes dadurch abschätzen, daß zu der Zahl die Anzahl Jahre hinzugefügt werden, die die Tanne benötigte, um eine Wuchshöhe von etwa 17 Metern zu erreichen. Bei einem durchschnittlichen jährlichen Wuchs von ca. 30 cm sind dies etwa 50 Jahre.

Zum Sortieren von Nadelholz mit den hier verwendeten Abmessungen (Zopfdurchmesser von mehr als 70 cm) gibt es keine Sortierregeln. Die Sortierregeln für Rundholz aus Nadelholz konnten hier nicht gelten, da alle Tannenstämme in der Stammachse über die ganze Länge zu Halbhölzern aufgetrennt wurden. Daher konnte das Sortieren nur in Anlehnung an DIN 4074 (Festigkeitssortierung von Nadelschnittholz), durchgeführt werden. Nadelschnittholz hat üblicherweise maximale Querschnittshöhen von 30 cm. Wegen der wesentlich größeren Abmessungen der hier verwendeten halben Rundholzquerschnitte ist davon auszugehen, daß die in den Sortierregeln der DIN 4074 angesetzten Werte, wie z.B. maximaler Astdurchmesser, viel zu klein sind und damit weit auf der sicheren Seite liegen. Zur Berücksichtigung dieses durch die Stammgröße bedingten Effektes wurden abweichend von DIN 4074 Astdurchmesser bis etwa 9 cm zugelassen. Diese und das Abweichen des Faserverlaufs von der Richtung der Stammachse wurden mit Hilfe eines Meßstabes visuell bestimmt.

Folgende Kriterien waren für die Sortierung maßgeblich:
- Astdurchmesser nicht über 90 mm. Es wurde jedoch sichergestellt, daß die zulässigen Zugspannungen im Bereich von Einzelästen mit Durchmesser > 60 mm (Bild 3) nur zu $2/3$ ausgenutzt werden.
- Jahresringbreite bis 4 mm. Diese Breite wird von 2 Stämmen geringfügig überschritten.
- Faserneigung bis 70 mm/m. Dieser Wert wird von einem Stamm geringfügig überschritten.
- Risse als radiale Schwindrisse zulässig. Blitz- oder Frostrisse und Ringschäle wurden bei keinem Stamm beobachtet.
- Verfärbungen zulässig.
- Druckholz bis zu $1/5$ des Querschnitts zulässig.
- Insektenfraß bei keinem Stamm beobachtet.
- Mistelbefall bei keinem Stamm beobachtet.
- Krümmung bis 5 mm / 2 m: Krümmung konnte noch nicht festgestellt werden, da die Stämme noch frisch waren. Im eingebauten Zustand wird die Krümmung behindert und wird zu Schwindrissen führen.

Über diese Sortierkriterien hinaus ist eine Mindestdarrdichte von 320 kg/m³ eingehalten.

Das Auftrennen der Tannenstämme zu Halbhölzern hatte nicht nur den Vorteil, daß dadurch eine schonende Trocknung der Stämme möglich und ihre Verbindung mit den übrigen Bauteilen der Konstruktion erleichtert wurde. Ein zusätzlicher Vorteil bestand darin, daß dadurch die Markröhre freigelegt war und damit mögliche biologische oder wachstumsbedingte Schädigungen in Form von Kernfäule, Rindeneinwüchsen oder überwachsenen Faulästen sofort erkennbar waren.

Bei Biegeprüfungen einer Stichprobe von 78 Kanthölzern, die aus den Restabschnitten der Tannenstämme gesägt wurden, konnten im wesentlichen die in Bild 4 gezeigten Bruchursachen festgestellt werden.

Es ist davon auszugehen, daß diese Kanthölzer geringere Festigkeiten als großvolumiges Rundholz besitzen, bei dem sich Beanspruchungen von lokalen Schwachzonen, z. B. Zonen mit Abweichungen der Fasern vom kantenparallelen Verlauf, auf großvolumige ungestörte Zonen verteilen können.

The typical material values contained in building standards for softwood were not fully applicable in providing proof of the load-bearing capacity of the canopy towers (consisting of four silver fir stems). Additional investigations were, therefore, necessary. The dimensions of the fir stems used for the towers differ considerably from those of softwood timbers normally used in construction work. As a first step, the timber density, knot sizes and deviations of fibre growth from the axis of the stem had to be determined for all the trunks used. The measurements were then checked to see whether they were comparable with equivalent known values for standard softwood members used in building, and whether a classification of the fir stems on the basis of cut softwood would be permissible. Since this grading into different classes determines the load-bearing capacity that may be assumed for the various members, the strength and rigidity of the fir stems were determined in failure tests in order to establish a relationship between the grading class and the bending strength and bending stiffness of the samples.

To determine the density of the wood, 10–15 cm thick cross-sections were sawn

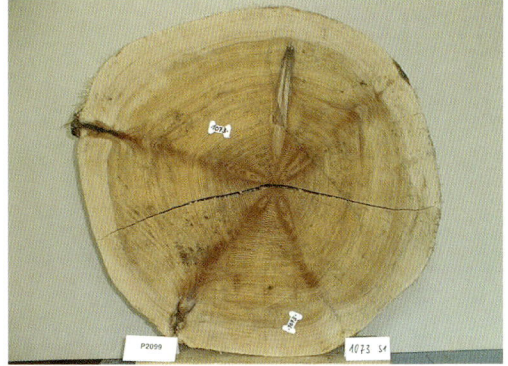

1 Stammscheiben / Stem cross-sections
a Stammscheibe mit 83 Jahrringen, Breite je 4,8 mm
a Section cut from stem with 83 annual rings, each 4.8 mm wide

b Stammscheibe mit 238 Jahrringen, Breite je 0,8 mm
b Section cut from stem with 238 annual rings, each 0.8 mm wide

2 Schrägfasrigkeit infolge Drehwuchs / Non-vertical fibre growth as a result of twisted grain

from the heads of 51 of the fir stems (ill. 1), which had been cut to about 17 m in length. The cross-sections had a diameter of between 66 and 86 cm. The frequency and distribution of the annual rings is shown in ill. 1. The age of a tree can be estimated from the number of rings in conjunction with known values for the number of years a fir tree requires to reach a height of 17 m. With an average annual growth of approximately 30 cm, this means a period of roughly 50 years.

No rules exist for the classification of softwood of the present dimensions (with a diameter of more than 70 cm at the head). The rules of classification for round softwood stems did not apply in this case anyway, since all the fir stems were cut in two along the longitudinal axis. DIN 4074 (the German Standard for grading cut softwood according to strength) could, therefore, be taken only as a reference value for classification. As a rule, squared, sawn softwood usually has a maximum cross-sectional depth of 30 cm. In view of the much larger dimensions of the half-round cross-sections used here, one may assume that the values given in the grading rules of the German Standard mentioned above (e.g. the maximum diameter of knots), were far too low and thus lay well on the side of safety. In order to take account of these factors – reflecting the size of the stems – knot diameters of up to about 90 mm were allowed (at variance to DIN 4074).

These and the amount of deviation of the fibre direction from the longitudinal axis of the stem were determined visually with the help of a measuring rod.

The following criteria were applied in classifying the stems:
- Maximum knot diameter: 90 mm. Care was taken to ensure that the permissible tension stresses in the area of individual knots with a diameter > 60 mm (ill. 3) were exploited to only $2/3$.
- Maximum annual ring width: 4 mm. In two stems, this width was minimally exceeded.
- Maximum fibre deviation: 70 mm per metre. In one stem, this value was minimally exceeded.
- Radial shrinkage cracks allowed. No cracks caused by lightning, frost or shakes were observed in any of the stems.
- Discoloration permissible.
- Compression wood up to $1/5$ of the cross-section permissible.
- No insect damage observed in any of the stems.
- No misteltoe damage observed in any of the stems.
- Maximum deviation in line of growth: 5 mm in 2 metres; no deviations were observed, since the stems were still freshly cut. After assembly, deflection will be prevented, which will lead to shrinkage cracks.

In addition to these criteria for classification, a minimum density of 320 kg/m³ after seasoning was maintained.

Cutting the fir stems into two semicircular halves had two advantages. It ensured a careful form of seasoning and facilitated an easier connection of the stems to other elements of the structure. It also meant that the medullary tubes were exposed, allowing potential forms of biological or growth-related damage – heart rot, ingrowing bark or overgrown unsound knots – to be recognized immediately.

In bending tests carried out on 78 squared specimens taken from cut-off lengths of the fir stems, the causes of the cracking shown in ill. 4 were determined to a large extent.

One may assume that the squared timbers used as test specimens have a lower strength than large-volume round stems, where the loading on local weak zones (e.g. zones where fibre orientation deviates from the axis line) is distributed over larger undamaged sections of the stem.

3 Astdurchmesser größer als 60 mm / Knot diameter exceeding 60 mm

4 Biegeprüfung / Bending tests
a Bruch infolge eines Astes
a Failure caused by a knot

b Bruch infolge von Schrägfasrigkeit
b Failure caused by non-vertical fibre growth

Tragwerksplanung

Structural Engineering

Julius Natterer, Norbert Burger,
Alan Müller, Johannes Natterer

Jeder einzelne Schirm besteht aus mehreren Baugruppen: vier Schalenflächen, vier auskragenden Trägern (Kragträgern), einer zentralen Stahlkonstruktion (Stahlpyramide, eigentlich ein Stahlpyramidenstumpf) und der Turmkonstruktion.

Die Schalenflächen weisen eine doppelte, gegensinnige Krümmung auf (Sattelflächen). Sie übertragen die Belastungen aus Eigengewicht, Schnee und Wind auf die Randglieder und die Stahlpyramide. Die Lastabtragung erfolgt dabei sowohl durch Schalentragwirkung (d.h. die Belastungen werden über Zug- und Druckkräfte in den Schalen abgetragen) als auch durch Biegetragwirkung.

Der über knapp 19 m frei ausladende Kragträger übernimmt die Lasten aus den Randträgern (Übertragung an der Spitze) und aus der Schale (Übertragung kontinuierlich am Untergurt). Der Untergurt des Kragträgers folgt der Krümmung des Schalenrandes und ist im äußeren Drittel mit dem Obergurt zusammengefaßt. Zur Schirmmitte hin nimmt die Bauhöhe entsprechend der Beanspruchung zu.

Die vier Kragträger sind an der Stahlpyramide angehängt. Durch die zentrale Stahlkonstruktion werden somit alle Kräfte zwischen den einzelnen Baugruppen der Schirmkonstruktion im Dachbereich übergeleitet und auf die Turmkonstruktion umgelenkt.

Über die Turmkonstruktion aus vier zusammengesetzten Rundholzstützen und dreiecksförmigen Aussteifungsscheiben werden alle vertikalen und horizontalen Lasten auf die Gründung übertragen. Die zentrale Stahlkonstruktion ist an den vier Stützenköpfen gelenkig mit den Stützen verbunden, so daß nur Normal- und Querkräfte übertragen werden. Die Aussteifungsscheiben übernehmen die gesamten horizontalen Belastungen aus Windbeanspruchung und unplanmäßiger Schiefstellung der Konstruktion.

Die Auflagerkräfte werden an den Stützenfußpunkten über Stahlfüße auf die Fundamente übertragen. Die Gründung besteht aus 4 vertikalen Großbohrpfählen Ø 1,20 m mit Längen von 10 bis 15 m. Die Pfahlköpfe sind durch einen Stahlbetonkranz verbunden, durch den die Spreizkräfte der Konstruktion ausgeglichen werden und durch den eine gemeinsame Lastabtragung der resultierenden Horizontalkräfte auf die Großbohrpfähle erreicht wird.

Die gesamte Konstruktion ist von einer Kunststoff-Dachhaut überspannt, die im Abstand von ca. 5 cm über der oberen Schalungslage »schwebt«. Damit die Dachhaut nicht flattert, ist sie gegen die Schalenränder vorgespannt.

Die Membranspannungen aus den Belastungen und aus der erforderlichen Vorspannung werden an den Schalenrändern auf die Schalungsebenen übertragen (zur Membrane siehe S. 30ff).

Lasteinwirkungen

Aus der Lage des Bauwerks ergibt sich eine Regelschneelast von $s_0 = 0{,}75$ kN/m². Die genauere Verteilung der Schneelasten konnte erst mit den Windkanalversuchen bestimmt werden. Dabei ergab sich eine größere Lastdifferenz zwischen benachbarten Schalen als die zunächst angenommene (70% anstatt 50% der maximalen Schneelast).

Eine abhebende Windwirkung, wie sie in Anlehnung an DIN 1055 zu erwarten wäre, war den Ergebnissen der Windkanalversuche in dieser Größenordnung nicht zu entnehmen. Hier zeigte sich dagegen eine stark belastende Wirkung nach unten. Verursacht wird dies durch die besondere Form der Dachfläche, auf deren Unterseite bei Anströmung Unterdruck entsteht, der sie nach unten zieht. Durch die Windkanalversuche ergaben sich somit deutlich veränderte Verhältnisse (siehe S. 52).

Die Rippenschale

Die Rippenschale ist als Brettstapelkonstruktion mit sich rechtwinklig kreuzenden Rippen ausgeführt und überdeckt eine Grundfläche von ca. 19 m x 19 m. Die vertikalen Abstände zwischen dem Tiefpunkt und den Hochpunkten betragen 6,0 m. Die Schale hat ein Gewicht von 37 t.

Die Rippenanordnung orientiert sich an den auftretenden Kräften in der Schale. Die Brettrippen weisen Querschnitte von 16/24 cm bis 16/30 cm auf und sind aus 8 bis 10 Brettlagen mit einem Querschnitt von 30/160 mm aufgebaut. Die einzelnen Brettlagen sind nachgiebig durch Verschrauben miteinander verbunden. In höher belasteten Teilbereichen sowie bei den Rippenanschlüssen an die Randglieder wurden sie verleimt. Die Brettlagen laufen an den Rippenkreuzungspunkten wechselseitig durch. Das jeweils quer verlaufende Brett ist als Füllbrett eingesetzt. In der Mitte der Kreuzungspunkte ist jeweils ein Paßbolzen angeordnet. Infolge des lagenweisen Aufbaus der Rippen ist die Herstellung der Rippenkreuzungspunkte einfach.

1

Luftaufnahmen des Baufortschritts / Aerial views showing progress of construction

1 Fundamente gegossen / Foundations laid

2 Erste Kragträger mit Dachschalen / Assembly of first cantilevered trusses with roof shells

3 Auslegen der Dachmembrane / Laying roof membrane

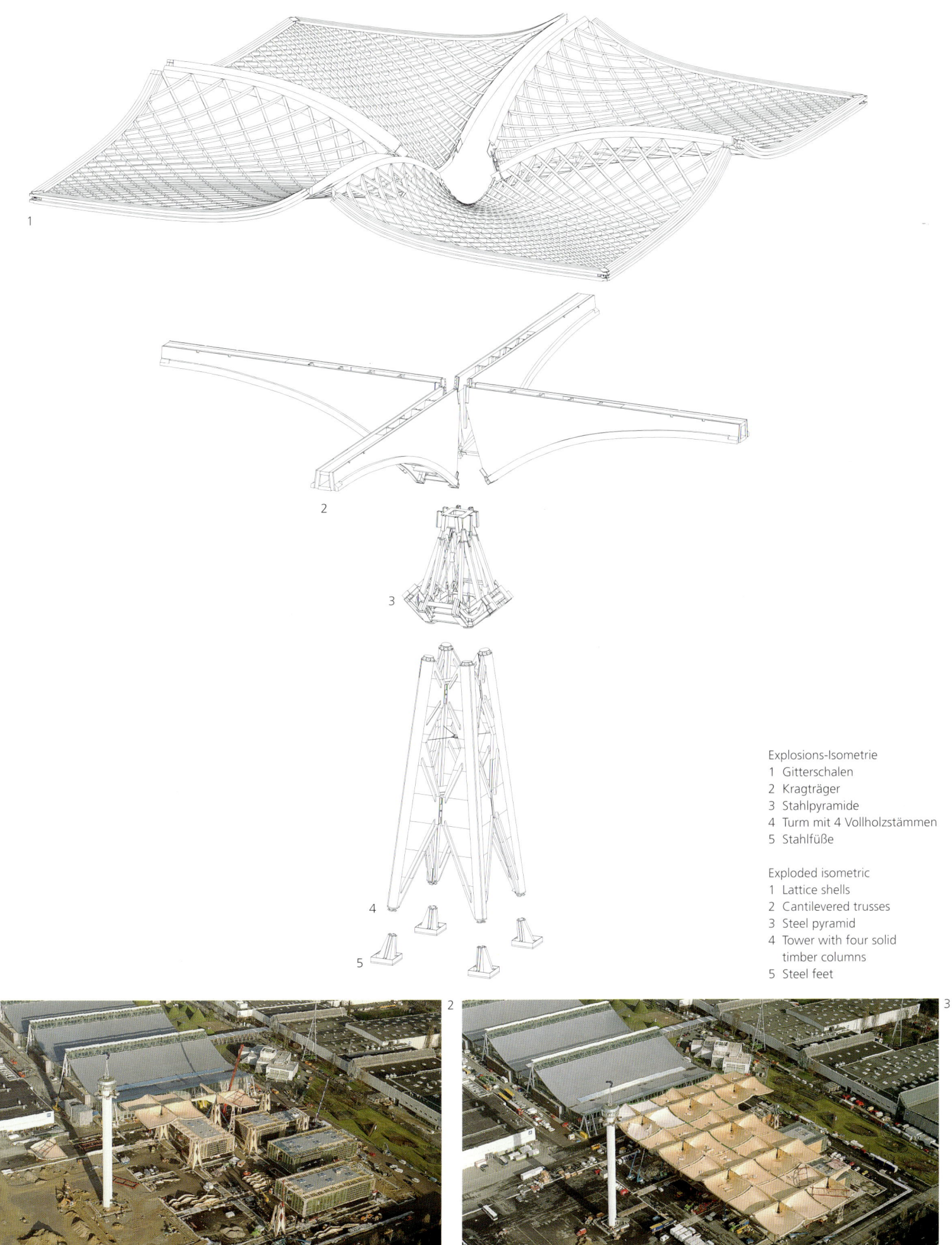

Explosions-Isometrie
1 Gitterschalen
2 Kragträger
3 Stahlpyramide
4 Turm mit 4 Vollholzstämmen
5 Stahlfüße

Exploded isometric
1 Lattice shells
2 Cantilevered trusses
3 Steel pyramid
4 Tower with four solid timber columns
5 Steel feet

Halbstämme nach Bearbeiten durch CNC gesteuerte Anlage / Semicircular logs after processing with CNC-programmed plant

Herstellen des Stützenanschlusses: vorne schwarze Dübel System BVD, oben silberfarbige Stabdübel / Insertion of dowels; at front: black dowels, BVD system, to rear: silver-coloured dowels

Zusammenfügen der zwei Stammhälften / Joining two halves of trunk together

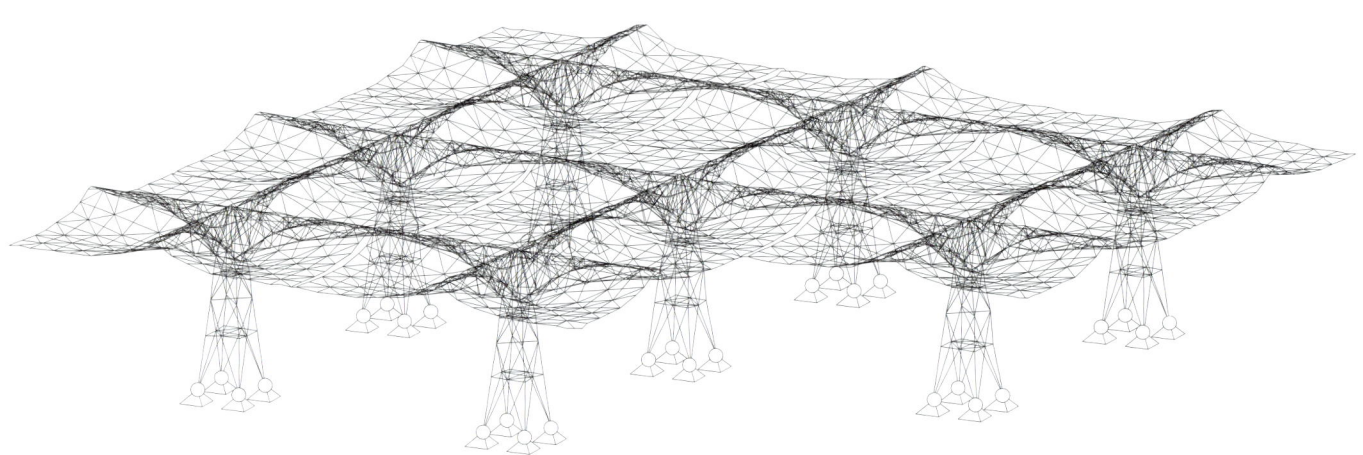

Zeichnungen / Drawings: IEZ

Isometrie und Dachaufsicht des statischen Rechenmodells für acht gekoppelte Schirme / Isometric and diagram of top of roof: calculation model for eight connected canopies

Die Rippenschale wird durch zwei unter 45° zu den Rippen verlaufende Brettlagen ausgesteift. Die einzelnen Bretter haben einen Querschnitt von 24/100 mm und sind im Abstand von 10 cm verlegt. Im Kehlbereich sind die Brettlagen wegen der hohen Kräfte und der starken Krümmungen durch eine in mehreren Schichten verleimte BFU-Schale ersetzt.

Um das dynamische Verhalten der Konstruktion nicht ungünstig zu beeinflussen (höhere Eigenfrequenzen infolge höherer Steifigkeiten und geringeres Dämpfungsmaß infolge steifer Verbindung), waren die verleimten Bereiche auf ein Mindestmaß zu beschränken.

Der Kragträger

Der Kragträger hat Abmessungen von ca. 19 m Länge, 2,9 m Breite und maximal 7,0 m Höhe. Er besteht aus zwei geneigten Teilquerschnitten mit veränderlicher Bauhöhe.

Die geraden Obergurte mit Querschnittsabmessungen von 22/100 cm bestehen aus Brettschichtholz BS 14 mit homogenem Aufbau. Die Untergurte sind entsprechend der Rippenschale gekrümmt und weisen mit 22/110 cm bis ca. 22/145 cm einen veränderlichen Querschnitt auf (BS 14 inhomogen). Im Bereich mit großer Trägerhöhe verlaufen Ober- und Untergurt getrennt und sind über einen Vollwandsteg aus zwei schraubpreßverleimten Schichten Furnierschichtholz (FSH, 2 x 33 mm) verbunden.

Laserstrahl zum Steuern der CNC-Anlage

Laser beam control of CNC plant

Blick auf die Werkzeugebene der CNC-Anlage

View of CNC tooling plant

▷ Turmkonstruktion aufgestellt

▷ Tower structure after erection

Im äußeren Drittel ist der Kragträger als Kastenquerschnitt aus Brettschichtholz ausgeführt und trägt die Kräfte aus den Randträgern wegen der geringen Bauhöhe über Biegung ab. Aus ungleicher Belastung der Schalen treten unterschiedliche Beanspruchungen in den Randträgern auf, die eine zusätzliche Torsionsbeanspruchung im Kragträger verursachen. Durch die Ausführung als geschlossener BSH-Kastenquerschnitt konnte in diesem Bereich eine große Biege- und Torsionssteifigkeit erreicht werden.

Der Anschluß der Gurtquerschnitte an die Stahlpyramide wurde mit Übergangsstücken als geschraubte Stahlbauverbindung ausgeführt.

Die Stahlpyramide

Die Stahlpyramide ist der zentrale Verbindungsknoten der Konstruktion, über den alle Kräfte aus den Rippenschalen und Kragträgern eingeleitet und auf die Turmkonstruktion umgelenkt werden. Sie hat Grundrißmaße von ca. 5,5 m x 5,5 m, eine Höhe von ca. 7,0 m und besteht aus je einem steifen oberen und unteren Kranz. Diese Kränze sind mit rechteckigen Hohlprofilen verbunden. Die Neigungen der Profile und Stege nehmen die Achsen der Turmkonstruktion und des Kragträgers auf.

Am unteren Kranz sind große Auflagerstege für den Untergurtanschluß angeschweißt. Das gekrümmte Widerlager für den Anschluß des Kehlbereiches der Schale stützt sich über Druckstreben auf die Auflagerstege und den unteren Kranz ab. Zum gleichmäßigeren Verteilen der Rippendruckkräfte ist vor der mittleren Druckstrebe zwischen dem gekrümmten Widerlager und dem gekrümmten Kehlstahlteil der Schale eine Tellerfeder angeordnet. Hohe Druckkräfte in den zentralen Längsrippen werden so gleichmäßiger auf die Längsrippen im Kehlbereich verteilt.

Die Aussteifung erfolgt über Auskreuzungen in Rundstahl, die vorgespannt sind, um Verformungen zu minimieren.

Die schweißtechnische Herstellung der Stahlpyramide stellte hohe Anforderungen und verlangte sowohl bei der Planung und als auch bei der Ausführung höchste Genauigkeit und Präzision.

Die Turmkonstruktion

Die Stützen mit Durchmessern von mind. 68 cm am unteren und bis zu ca. 110 cm am oberen Ende bestehen aus einem Baum, der der Länge nach mittig in zwei annähernd gleiche Halbrundstämme aufgetrennt wurde. Die Halbrundstämme sind mit einem Abstand von 63 mm über Zwischenhölzer im Abstand von 50 cm bis 75 cm nachgiebig mit Einpreßdübeln und Bolzen verbunden.

Durch den kleineren Achsabstand am Stützenkopf treten hier größere Beanspruchungen in der Stütze auf als am Stützenfuß mit deutlich größerem Achsabstand. Die Stützen sind deshalb, der Größe der Beanspruchung folgend, entgegen der natürlichen Wuchsrichtung mit dem kleineren Durchmesser unten eingebaut (d. h. der Baum »steht auf dem Kopf«). Am Stahlfuß und an der Stahlpyramide sind die Stützen über ein Stahlübergangsstück mit vorgespannten Schrauben angeschlossen. Mit der Holzstütze sind die Übergangsstücke über 8 Ankerkörper des Verbindungssystems Bertsche-BVD verbunden.

Der Turm ist durch Diagonalen in Brettschichtholz mit beidseitiger Beplankung aus Furnierschichtholz-Platten ausgesteift. Auch diese Aussteifungsscheiben sind mit den Stützen kontinuierlich über Einpreßdübel verbunden. Zum Ausgleich der Schwindverformungen werden die Bolzen der Einpreßdübel durch eingebaute Tellerfedern auf Spannung gehalten. Die Aussteifungsdiagonalen sind über Stahldübelverbindungen nachgiebig an die Stützen angeschlossen.

Ein seitliches Umklappen der Aussteifungsscheiben und gegenseitiges Verdrehen der Stützen wird durch zwei horizontale Schottenebenen und horizontale Stahlzugstäbe zwischen den zentralen Verbindungsknoten der Scheiben verhindert. Weitere horizontale Stahlzugstäbe verbinden je zwei gegenüberliegende Stützen auf Höhe der Verbindungsknoten der Aussteifungsscheiben mit den Stützen.

Horizontalschnitte der Turmkonstruktion in verschiedenen Höhen   Maßstab 1:125

Horizontal sections through tower structure at various heights   scale 1:125

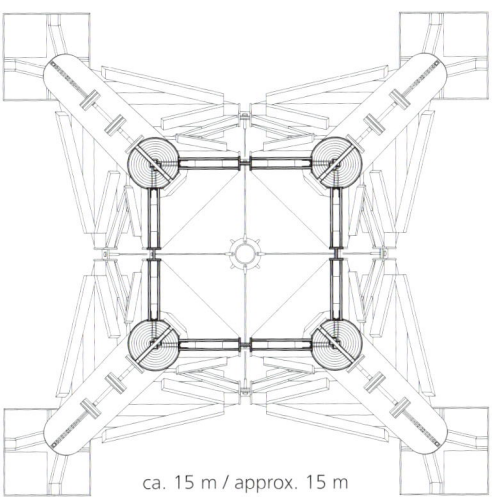

ca. 15 m / approx. 15 m

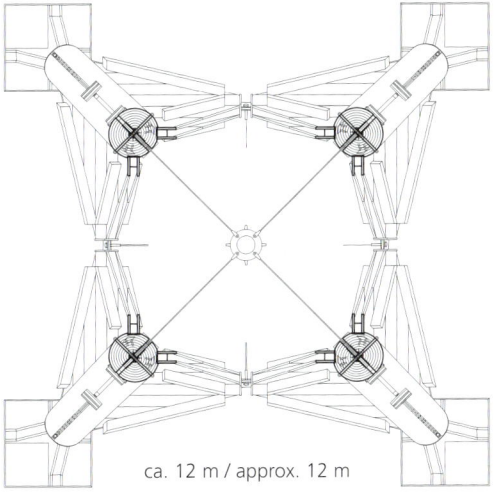

ca. 12 m / approx. 12 m

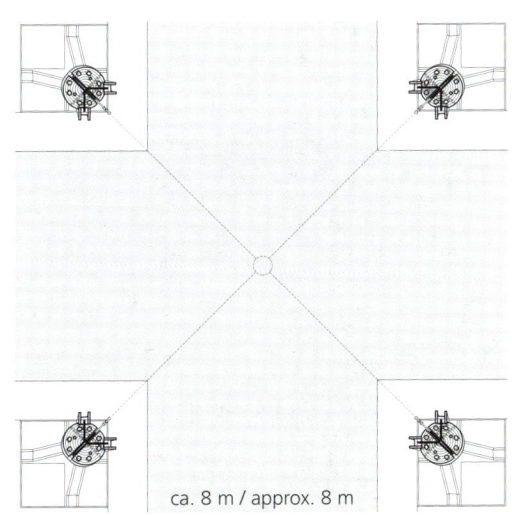

ca. 8 m / approx. 8 m

Montage des Regen-Fallrohrs

Assembly of rainwater pipe

Regen-Fallrohr, ausgeformtes unteres Ende

Rainwater pipe with tapered lower end

Fußpunkt des Turms mit Ende des Regenfallrohrs über einer »Gracht«
1 Stamm
2 Turm-Aussteifung
3 Stahlfuß
4 Ringfundament
5 Oberes Ende der Pfahlgründung
6 Wasseroberfläche der »Gracht«
7 Regenwasser-Fallrohr
8 Fallrohr-Abspannung

Foot of tower with end of rainwater pipe over a "gracht"
1 Tree-stem column
2 Tower bracing
3 Steel foot
4 Foundation ring
5 Head of pile foundations
6 Water level in "gracht"
7 Rainwater pipe
8 Pipe stays

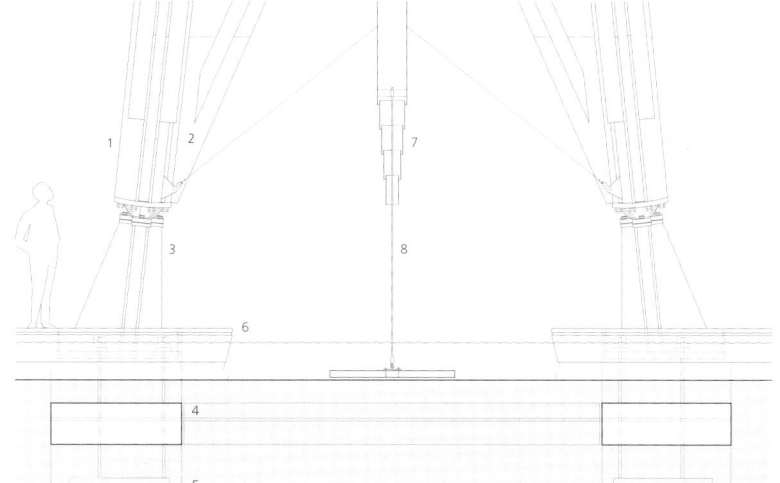

Each of the canopies consists of a series of constructional groups: four shell elements, four cantilevered trusses, a central steel structure (steel truncated pyramid) and the tower construction.

The shell elements are double curved in opposite directions (saddle surfaces) and transmit their loading, consisting of dead load, snow and wind loads, to the edge members and the steel pyramid. Load transmission occurs, therefore, via the load-bearing behaviour of the shells themselves (i.e. in the form of tension and compression forces in the shells) and through bending.

The nearly 19-metre-long freely cantilevered trusses bear the loads from the edge beams (transmission at the tips) and from the shell (continuous transmission along the lower chord). The lower chords of the cantilevered trusses follow the curvature of the edge of the shell and are connected to the upper chords along the outer third of their length. Towards the centre of the canopy, the depth of the construction increases with the loading actions.

The four cantilevered trusses are fixed to the steel pyramid. All loads borne by the individual elements of the canopy roof are ultimately transmitted to the central steel structure and from there to the tower. All vertical and horizontal loads are transmitted down to the foundations via the tower construction, which consists of four built-up round timber columns with triangular bracing members between them. The central steel structure is flexibly fixed to the heads of the four columns, so that only axial and shear forces are transmitted. The bracing slabs resist all horizontal loads resulting from wind action and from any unintended non-verticality of the structure itself.

The bearing forces at the base of the columns are conveyed via steel feet to the foundations, which consist of four vertical large bored piles 10–15 m long. The pile heads are connected by a reinforced concrete ring, which serves to equalize the different horizontal loads on the heads of the piles.

The entire roof construction is covered by a membrane that "floats" roughly 5 cm above the upper layer of boarding. To keep it taut and prevent it from flapping up and down, the roof skin is prestressed against the edges of the shells.

Membrane stresses resulting from loading and the necessary prestressing are transmitted to the edge members of the shells in the plane of the boarding (for details of membrane, see page 30ff).

Loading actions

Taking account of the location of the structure, an average snow load of $s_o = 0.75$ kN/m² has to be assumed. Only after carrying out wind-tunnel tests was it possible to determine the distribution of the snow loads more precisely. The outcome was a greater load difference between adjoining shells than at first assumed (70 per cent instead of 50 per cent of the maximum snow load).

The expected lifting effect of wind was not corroborated in the wind-tunnel tests to the extent indicated by German Standards (DIN 1055). On the contrary, the tests showed a heavy download effect resulting from the special form of the roof surface, which exhibits a similar aerodynamic behaviour to

Stahl-Kopfplatten mit Anschlußschwertern für die Aussteifungen / Steel head plates with connection fins for bracing

Kopfteil der Stahlpyramide: Anschluß für den Obergurt des Kragträgers

Head of steel pyramid with connection plates for top chord of cantilevered truss

Zusammenbau einer Stahlpyramide im Werk / Assembly of steel pyramid at works

that of an inverted aeroplane wing. Negative pressure occurs on the underside and draws the roof downwards. The wind-tunnel tests, therefore, revealed quite different conditions from those expected (see page 52).

Ribbed shells

The ribbed canopy shell is a stacked-plank structure with a lattice grid of ribs that intersect at right angles to each other. The individual elements cover an area roughly 19 x 19 m in extent. The difference in height between the lowest and highest points is 6.0 m. Each shell weighs 37 tonnes.

The layout of the ribs reflects the forces acting within the shell. The ribs consist of 8–10 stacked planks, each of which is 30/160 mm in cross-section. The overall cross-sections of the ribs range from 16/24 cm to 16/30 cm. The individual planks are compliantly screwed to each other. In zones subject to greater actions and at the junctions between the ribs and the edge members, the rib layers are glued together. Alternate plank layers continue across the points of intersection of the ribs at right angles to each other. The planks that do not continue across the intersection are cut off and butt jointed at the sides. A fitted bolt is fixed through the ribs at the centre of each intersection. The layered build-up of the ribs facilitates a simple form of construction at the points of intersection.

In order to avoid adverse effects on the dynamic behaviour of the structure (higher resonance frequencies resulting from greater stiffness, and reduced damping as a result of rigid connections), areas of glued bonding were reduced to a minimum.

Cantilevered trusses

The cantilevered trusses are roughly 19 m long and 2.9 m wide, with a maximum depth of 7.0 m. They consist of two curved and splayed cross-sections of varying structural depths.

The straight top chords, with cross-sectional dimensions of 22/100 cm, consist of glued laminated timber (BS 14) of homo-

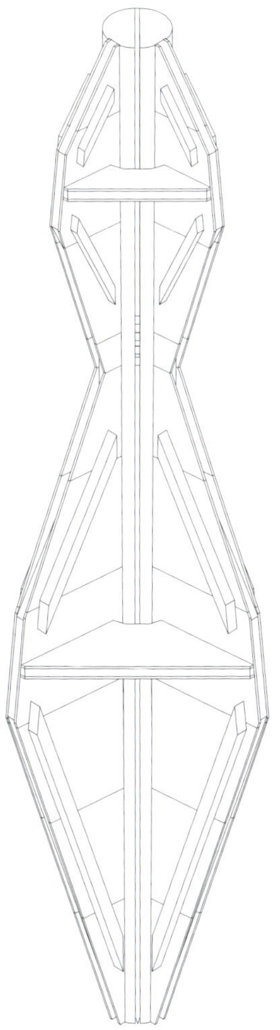

Holzteile / Timber components

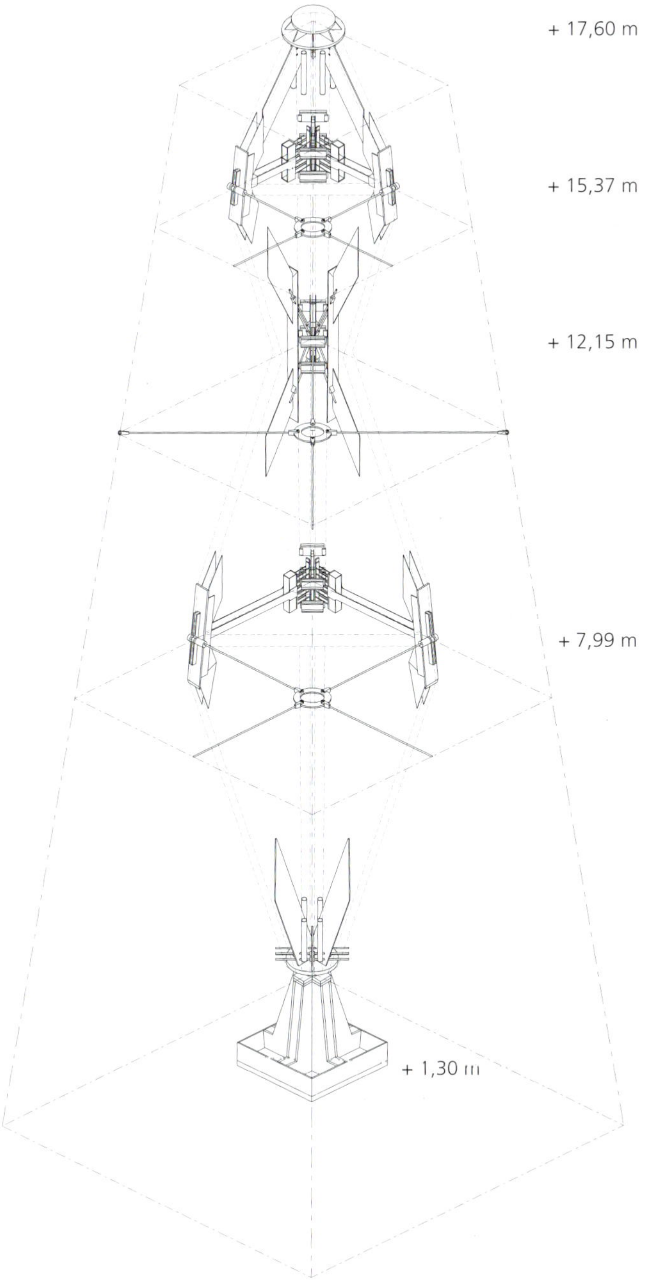

Holz- und Stahlteile am Turm / Timber and steel components in tower

Stahlteile / Steel components

+ 17,60 m
+ 15,37 m
+ 12,15 m
+ 7,99 m
+ 1,30 m

Unterer Ansatz der Stahlpyramide /
Lower seating of steel pyramid

geneous structure. The lower chords (BS 14, non-homogeneous) follow the curved line of the ribbed shells and have cross-sections varying from 22/110 cm to roughly 22/145 cm. Over the length where the trusses are at their deepest, the upper and lower chords diverge, but are connected by solid webs, consisting of two layers of screw-fixed and glued laminated timber sheeting (2 x 33 mm).

Over the outer third of their length, the cantilevered trusses are constructed as box sections in glued laminated timber. In view of their minimum depth here, they bear the loads from the edge beams as bending forces. Unequal loading in the shells results in different stresses in the edge beams, which, in turn, cause additional torsional actions in the cantilevered trusses. The closed, glued-laminated timber box section of the trusses ensures great bending and torsional stiffness in these zones. The chords are connected to the steel pyramid with interface members in the form of bolted steel connections.

Steel pyramids

The steel pyramid forms the central connecting node of the entire construction, through which all loads from the ribbed shells and cantilevered trusses are transmitted to the tower structure. The pyramid has base dimensions of roughly 5.5 x 5.5 m and a height of about 7.0 m. It consists of a rigid upper and lower ring connected by rectangular hollow sections. The angle of inclination of the sections is aligned with the axes of the tower construction and the cantilevered trusses.

Welded to the lower ring are large interface seatings for connecting the lower chords of the trusses. The curved abutment for the connection of the throat section of the shell is supported by struts bearing on the abutment members and the lower ring. To ensure a more even distribution of the compression actions from the ribs, a disc spring was inserted in front of the central strut between the curved seating and the curved steel throat member of the shell. As a result, great compression actions in the central longitudinal ribs are distributed more evenly over these ribs in the throat area.

The construction of the steel pyramids presented a great challenge in terms of welding technology and required the utmost precision, both in the planning and the execution of the work.

Tower construction

The columns, with diameters of at least 68 cm at the base and up to about 110 cm at the upper ends, are constructed from single tree stems cut lengthways into two virtually equal halves. The semicircular stems are compliantly joined together with dowelled and bolted connections and with distance pieces at 50–75 cm centres, leaving a space of 63 mm between the two halves.

The closer spacing of the column axes at the head of the tower means that greater loads are imposed on the columns here than at the feet, where the spacing is much greater. To reflect this, the columns were assembled contrary to their natural direction of growth – with the smaller diameter at the bottom. In other words, the tree stems actually stand on their heads. The columns are connected to the steel base and the steel pyramid by means of steel seatings with high-tensile bolts. The seating elements are fixed to the timber columns with eight anchor pieces from the Bertsche-BVD connecting system.

The tower is braced by diagonal glued-laminated timber members clad on both faces with laminated veneer timber boards. These bracing slabs are also fixed continuously to the columns with dowels. To counteract deformation caused by shrinkage, the dowels are held in tension by in-built disc springs. The diagonal bracing members are compliantly fixed to the columns with steel dowel-type fasteners.

Any pivoting action in the bracing slabs or reciprocal twisting of the columns is prevented by two horizontal cross-layers or "bulkheads" and by horizontal tension rods between the central connection nodes of the bracing slabs. Further horizontal steel tension rods connect pairs of opposite columns at the level of the connecting nodes between the bracing slabs and the columns.

Dreidimensionale Rechenmodelle

Three-Dimensional Calculation Models

Die ingenieurmäßige Planung und Umsetzung von Bauwerken ohne den durchgängigen Einsatz elektronischer Hilfsmittel ist heute nicht mehr vorstellbar. Dies gilt – neben der Werkstattplanung – vor allem für die Tragwerksplanung, selbst wenn nahezu alle Bauwerke ebenso mit traditionellen »Hand«-Rechenmethoden bemessen werden könnten. Um die Struktur berechenbar zu machen, sind jedoch dort je nach Art und Komplexität der Tragwerke mehr oder weniger starke Vereinfachungen zu treffen. Als Folge ergeben sich in aller Regel unwirtschaftlichere Konstruktionen und ein unterschiedliches Sicherheitsniveau.

Auch die Konstruktion des EXPO-Daches erscheint ohne computergestützte Methoden nicht unmöglich, wie Kontrollrechnungen der Schale gezeigt haben. Derart umfangreiche Untersuchungen mit Variation verschiedener Parameter, wie sie im Verlauf der Planung beim EXPO-Dach mit gesamtheitlicher Optimierung der Querschnitte und des Tragverhaltens vorgenommen wurden, sind jedoch ohne sehr hohe Rechnerleistung nicht denkbar.

Die Größe und Komplexität der Konstruktion machte eine Untersuchung des Tragverhaltens anhand mehrerer Rechenmodelle erforderlich (Schale, Schirm, Gruppierung u. a.). Die wesentlichen Bestandteile der Konstruktion sind stabförmige Bauteile. Sowohl die Gesamtstruktur als auch die Teilstrukturen wurden daher als Stabwerk idealisiert, d. h. alle Bauteile wurden in Form einzelner Stäbe oder Stabzüge modelliert.

Eine wichtige Voraussetzung für die optimierte Berechnung und Bemessung einer solch komplexen Struktur wie der des EXPO-Daches ist die rasante Entwicklung der Informatik in den letzten zehn Jahren. Wegen der dennoch hohen Rechenlaufzeiten wurden für Parameterstudien häufig parallele Rechenläufe an mehreren Rechnern durchgeführt.

Blick von unten in die Turmkonstruktion, Montagezustand

View from below of tower structure during assembly

Die wesentlichen Rechenmodelle zur Berechnung eines gesamten Schirmes waren: die Schirmstruktur mit den Schalenflächen und Kragträgern, die Stahlpyramide und die Turmkonstruktion.

Das Rechenmodell der Schirmkonstruktion setzt sich insgesamt aus etwa 2500 räumlich angeordneten Knotenpunkten und nahezu 9000 Einzelstäben zusammen, die alle in ihrer Geometrie, den Materialeigenschaften, der Lage und den Knotenverbindungen definiert sind. Am Übergang vom Kragträger zur Stahlpyramide sind feste Auflager eingefügt.

Die Berechnungen für die Turmkonstruktion wurden aus terminlichen Gründen vorgezogen. Die Belastungen auf dem Schirm wurden über Gleichgewichtsgleichungen auf entsprechende Kräfte am Turmkopf umgerechnet, die als Belastungen auf die Turmkonstruktion angesetzt wurden. Damit konnten die Belastungen auf die Fundamente bereits zu einem sehr frühen Zeitpunkt bestimmt werden.

Die Nachgiebigkeit von Anschlüssen wurde in den Rechenmodellen durch entsprechende Reduzierung der Querschnittswerte berücksichtigt.

Um die Schnittkräfte und Verformungen zu überprüfen, wurden in einem Vergleichsmodell die Rechenmodelle der Turmkonstruktion und der Stahlpyramide zusammengesetzt und berechnet. Auch die Schnittstelle zur Schirmstruktur wurde auf Verträglichkeit der Verformungen und auf Übereinstimmung der Schnittgrößen untersucht. Die Überprüfung des Gesamtsystems war aufgrund des komplexen Zusammenwirkens der Schalen und der Kragträger zwingend erforderlich.

Die Rechenmodelle der Schirmkonstruktion gehen bis an die Leistungsgrenze der heute verfügbaren Ausrüstung. Das verwendete Rechenprogramm wurde mehrfach modifiziert.

Die Entwicklung des EXPO-Daches beschränkt sich so nicht nur auf die Tragwerksplanung, sondern führte – parallel dazu – auch zu einer deutlichen Verbesserung des Statikprogrammes.

The engineering planning and the execution of structures of this kind would be inconceivable today without the use of electronic aids. This applies not only to the workshop phase of the work, but above all to the structural planning, even though almost all structures could be calculated "by hand", using traditional methods. Depending on the nature and complexity of a structure, however, a number of greater or smaller simplifications would have to be made to facilitate the calculations. In general, this would mean that the structures would be less economical in their construction, and the safety factors would vary.

Probably the structural calculation of the EXPO roof would also have been possible without computer-aided methods, as control calculations for the shell revealed. In the course of planning this roof, however, extensive investigations were undertaken in which the parameters were varied in order to achieve an overall optimization of the cross-sections and the load-bearing behaviour. These calcu-

Komplexe Gesamtgeometrie im Turmbereich, übereinander projiziert /
Complex overall geometry in tower areas; elements superimposed

Lehrgerüst mit Fertigung der Gitterschale /
Centring with construction of lattice-grid shell

Fertigung der Gitterschale /
Construction of lattice-grid shell

lations would certainly not have been conceivable without intense computer input.

In view of the size and complexity of the structure, it was necessary to investigate its load-bearing behaviour, using a number of calculation models. The main structural components are linear framed members. Both the overall structure and its individual subsections were, therefore, idealized as a linear framework; i.e. all components were treated as individual linear members or bar chains in the model.

The rapid development of information technology over the past ten years is one of the most important factors that have made an optimized calculation and dimensioning of complex structures like the EXPO roof possible. In conducting parameter studies, parallel calculations were often made on a number of computers.

Rigid abutments were foreseen at the junctions between the cantilevered trusses and the steel pyramid. The flexibility of the tower construction was modelled using appropriate spring bearings. In order to check the internal forces and the deformation, the calculating models for the tower structure and the steel pyramid were combined and calculated in a comparative model. An investigation of the entire system was essential in view of the complex reciprocal action between the shells and the cantilevered trusses.

The calculation model comprises some 2,500 nodes in a three-dimensional layout and almost 9,000 individual bars or linear members, all of which were defined in terms of their geometry, material properties, position and node connections.

The calculations for the tower structure were carried out using a separate model. The loads on the canopy were converted by means of equilibrium equations into corresponding forces acting on the head of the tower and were calculated as loads on the tower structure itself. As a result, it was possible to determine the loads on the foundations at a very early stage.

The calculation models used for the canopy construction went to the limits of the equipment available today. The development of the EXPO roof involved not just the structural planning; parallel to this, a substantial improvement was also achieved in the statical programme.

Schalengeometrie

Shell Geometry

Bei den bisher bekannt gewordenen Ausführungen von Schalen in Form hyperbolischer Paraboloide (HP-Schalen) entstehen gerade Randglieder. Eine Konstruktion mit gekrümmten Randgliedern ist in dieser Dimension neuartig.

Die vertikale Spreizung der Struktur (Höhenunterschiede zwischen den vier Schalenecken) hat einen wesentlichen Einfluß auf die auftretenden Kräfte und damit auf das Trag- und Verformungsverhalten der Schale.

Die Fläche soll möglichst an die Form einer Minimalfläche (auch Membranfläche) angenähert sein. Ihre Geometrie ergibt sich beispielsweise durch Aufspannen einer biegeweichen Fläche, z. B. einer Seifenhaut oder eines Seilnetzes, zwischen den als starr angenommenen Randgliedern.

In einer Membranfläche treten nur Normalspannungen, d. h. Zug- und Druckkräfte auf. Bei einer veränderten Belastung ändert sich die Form der Fläche, und an den Rändern treten Spannungskonzentrationen und höhere Kräfte auf.

Die Schalenfläche wird durch eine geschlossene mathematische Funktion beschrieben. Fixpunkte für die Definition sind die Koordinaten der vier Eckpunkte der Schale (freie Außenecke der Schale, zwei Hochpunkte der Schale an den Enden der Kragträger, Auflagerpunkt im Kehlbereich). Die Untergurte des Kragträgers werden durch eine Parabel 3. Ordnung mit Scheitelpunkt (horizontale Tangente) an der Spitze des Kragträgers, die Hauptdiagonale vom Kehlbereich bis zur freien Außenecke durch eine Parabel 2. Ordnung mit Scheitelpunkt an der freien Ecke beschrieben. Zwischen die Kragträgerachsen wurden Parabeln 2. Ordnung mit Scheitelpunkt auf der Hauptdiagonalen eingehängt. Die Achsen der Randträger ergeben sich durch den Schnitt mit vertikalen Ebenen durch die Spitze des Kragträgers und die freie Ecke der Schale.

Anhand der mathematisch definierten Schalenform wurden die Rippenachsen als geodätische Linien bestimmt. Die Abstände der Rippen wurde statischen Erfordernissen angepaßt. Sie variieren zwischen 38 cm im Bereich der Hauptdiagonale und 160 cm im Bereich der Schalenhochpunkte.

Um nach unten hängende Außenecken (»lahme Flügel«) zu vermeiden, ist beim Festlegen der Koordinaten für die ausgeführte Schale die vertikale Verformung durch eine Überhöhung von 15 cm im äußeren, frei auskragenden Teil der Schale berücksichtigt. Dies geschah nicht nur aus ästhetischen Gründen, sondern auch um eine mögliche Wassersackbildung zu vermeiden.

Traditional shell structures, in the form of hyperbolic paraboloids, are distinguished by straight, linear edge members. A structure on the present scale with curved edge members is something entirely new.

The vertical spreading of the structure (differences of height between the four corners of each shell) has a major influence on the loading that occurs and thus on the load-bearing behaviour and the deformation of the shell.

The surface should ideally approximate a minimum area (or membrane area) as far as possible. Its geometry may be derived, for example, from the form obtained by stretching a flexible, elastic skin, such as a soap bubble or a cable net, between the edge members, which are assumed to be rigid.

Only axial forces occur in a membrane surface: in other words, tension and compression loads. If the loading is varied, the surface form will change, and stress concentrations and higher loads will occur at the edges.

The shell surface can be described by means of a closed mathematical function. The co-ordinates of the four corners of the

Untersicht der fertig montierten Gitterschalen / Underside of lattice-grid shells after assembly

Fertige Gitterschale im Lehrgerüst / Completed lattice-grid shell on centring

shell serve as fixed points for its definition (the unsupported outer corner of the shell; the two high points of the shell at the ends of the cantilevered trusses; and the point of abutment at the "throat" next to the pyramid). The lower chords of the cantilevered trusses describe a parabola of the third order with its apex (horizontal tangent) at the outer tip of the truss. The main diagonals, from the throat to the unsupported outer corner of the shell, describe a parabola of the second order, with its the apex at the free outer end. Suspended between the axes of the cantilevered trusses and the axes of the edge beams are parabolas of the second order, the vertices of which lie along the main diagonals. The axes of the edge beams are defined by drawing a vertical section through the tip of the cantilevered truss and the outer unsupported corner of the shell.

The rib axes were determined as geodesic lines on the basis of the mathematically defined shell form. The rib spacings were calculated in accordance with structural needs. They vary from 38 cm in the area of the main diagonal to 160 cm at the crests of the shells.

In determining the co-ordinates for the shell construction, a banking was allowed for vertical deformation. The outer, unsupported tip of the shell was raised by 15 cm to prevent a sagging effect ("hanging wings") at the corner. This was done not only for aesthetic reasons, but to avoid the formation of water pockets.

Mathematische Entwicklung der Gitterschale / Mathematical development of lattice-grid shell

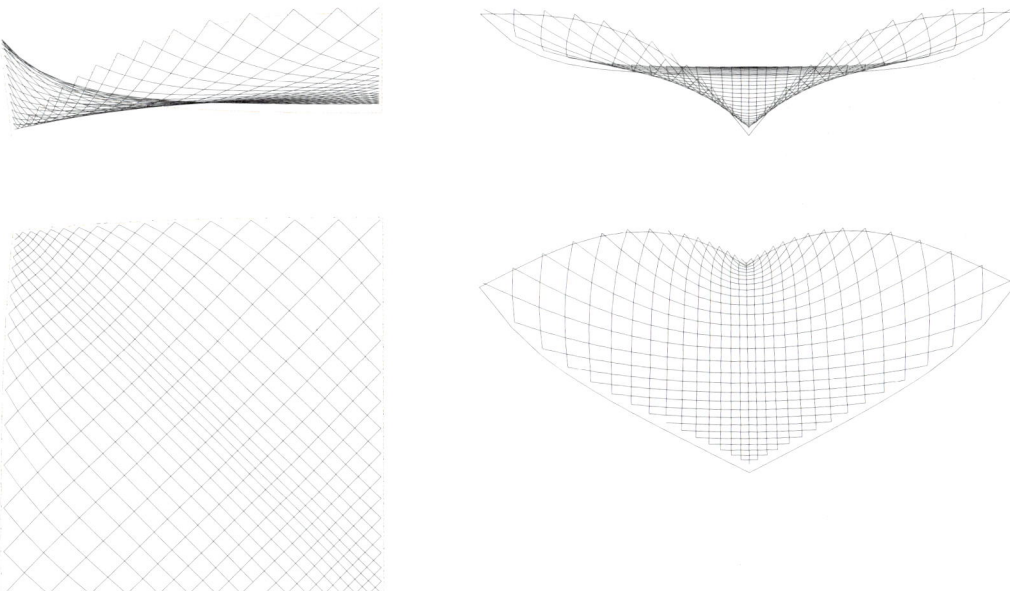

Dynamisches Verhalten der Dachkonstruktion

Dynamic Behaviour of Roof Construction

Heinrich Kreuzinger

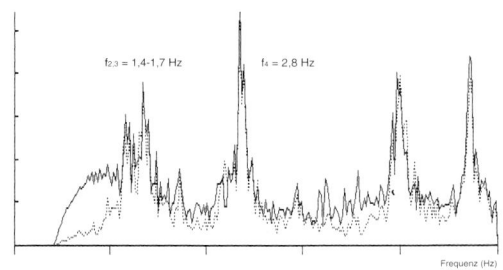

1 Spektrum der vertikalen Bewegung / Spectrum of vertical movement

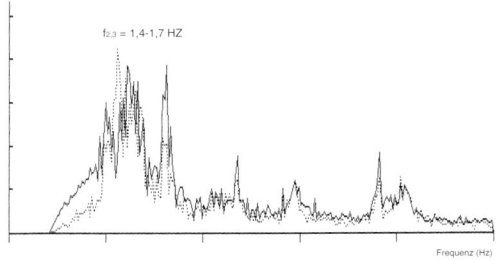

2 Spektrum der horizontalen Bewegung / Spectrum of horizontal movement

Bis auf das Eigengewicht einer Konstruktion sind alle anderen Einwirkungen mehr oder weniger zeitlich veränderlich. Die zeitliche Veränderung der Einwirkung und das Verformungsverhalten des Bauwerks hängt vom Zusammenwirken der zeitveränderlichen Eigenschaften beider ab. Wesentliche Kenngrößen sind der Rhythmus, die Frequenz der Anregung und die Eigenfrequenzen des Bauwerks. Die Eigenfrequenzen eines Bauwerks geben an, wie oft ein Bauwerk nach einer Anregung sich selbst überlassen hin- und her schwingt. Stimmt der Rhythmus der Anregung mit der Eigenfrequenz einer Eigenform des Bauwerks überein, so kann es zu großen Bewegungen, zu Resonanzschwingungen kommen.

Wind weht nicht immer gleichmäßig. Die Windeinwirkung auf Bauwerke ist über den Staudruck q von der Dichte ρ der Luft und dem Quadrat der Windgeschwindigkeit v abhängig. Der Einfluß der Bauwerksform wird rechnerisch durch einen Formbeiwert c erfaßt. Für übliche Bauwerke ist der Formfaktor bekannt. Die Windgeschwindigkeit selbst unterliegt starken Schwankungen. Die Einwirkung auf das Bauwerk ist dann noch mehr schwankend und kann ein Bauwerk zu Schwingungen anregen. Windböen zeigen dies anschaulich. Daneben kann selbst bei konstanter Windgeschwindigkeit eine Anregung durch Wirbelablösungen senkrecht zur Windgeschwindigkeit auftreten.

Der zeitliche Abstand von Böen bzw. die zeitlichen Schwankungen der Windeinwirkung wird von der Frequenz abhängig in Spektren angegeben. Über den Übertragungsmechanismus des Bauwerks kann das Spektrum der Bauwerksantwort erhalten werden. Für übliche Bauwerke ist dies bekannt und kann rechnerisch für ein bestimmtes Bauwerk nachvollzogen werden. Für das Dach aus den 10 Schirmen ist eine rein rechnerische Beurteilung schwierig: der Formbeiwert ist nicht bekannt, Form und Größe haben Auswirkungen. Aus Windkanalversuchen wurden aerodynamische Formbeiwerte für die Dachstruktur gewonnen (s. S. 52). Die Böigkeit des Windes und das dynamische Verhalten des Daches wurden dabei im Versuch abgebildet.

Zur Modellbildung waren die Eigenfrequenzen und die zugehörigen Eigenformen des Daches notwendig. Dazu wurden Berechnungen gemacht. Abb. 3 zeigt die Eigenschwingform eines Schirmes, bei der er sich über die Diagonale hin und her bewegt, Abb. 4 die Eigenschwingform, bei der diagonal gegenüberliegende Schirmfelder sich nach oben oder unten bewegen. Nach Fertigstellung erfolgte eine Schwingungsmessung am Schirm in der Süd-West-Ecke durch das Fachgebiet Holzbau der TUM.

Die Messung fand vor der Koppelung der einzelnen Schirme (s. S. 60) statt.

Eine mathematische Auswertung der gemessenen Beschleunigungswerte gibt vorherrschende Frequenzen an. Da zwei Aufnehmer in verschiedenen Richtungen und an verschiedenen Stellen der Dachkonstruktion angeordnet wurden, kann aus dem Vergleich auf die zur Frequenz gehörende Eigenform geschlossen werden. Abb. 1 zeigt ein gemitteltes Amplitudenspektrum der vertikalen Beschleunigung, Abb. 2 das der horizontalen Beschleunigung. Im Spektrum der horizontalen Beschleunigungen fehlt die Spitze bei etwa 2,8 Hz. Das erlaubt den Schluß, daß zu dieser Frequenzspitze in Abb. 1 eine Eigenform gehört, die keine horizontalen Bewegungsanteile hat, wie z. B. Schmetterlingsform nach Abb. 4.

Berechnungen und Windkanalversuche gaben die Bemessungsgrundlagen für die Dachkonstruktion. Messungen am fertigen Bauwerk bestätigen die berechneten dynamischen Kenngrößen des Systems.

With the exception of the self-weight of a structure, all influences on it are subject to change from time to time. The temporal changes that occur as a result of these influences and the deformation behaviour of a built structure will depend on their interaction over a period of time. The main quantifiable factors in this respect are the rhythm, the frequency of the activation, and the natural frequencies of the structure. The natural frequency defines how often a structure will vibrate backwards and forwards after stimulation if left to its own devices. If the rhythm of the activation coincides with the natural frequency of a mode of the structure, the outcome can be major movements and resonant vibration.

Wind does not always blow at a constant speed. Its effect on a building structure will depend on the back pressure of the density of the air, and on the square of the wind speed. The influence of the form of the structure itself can be calculated using a form factor. For common structures, the form factor will be known; but the wind speed is subject to great fluctuations. The effects of wind on the building, therefore, will be even more variable and can lead to vibrations in the structure. This is vividly demonstrated when a structure is subject to gusty winds. But even with constant wind speeds, the building can be activated by a succession of whirling currents at right angles to the direction of the wind.

The intervals between the gusts of wind or the fluctuations in wind activity over a period of time are defined in spectra according to the frequency. The spectrum for the reaction of the structure can be obtained by means of the transmission mechanism of a building. In the case of common forms of construction, this will be known and can be calculated for a given building. For the present roof, which consists of ten canopies, an evaluation based purely on calculations is problematic: the coefficient for the form is not known, and the form and size have an effect on the whole. Wind-tunnel tests were used to obtain aerodynamic coefficients for the form of the roof structure (see page 52). The gustiness of the wind and the dynamic behaviour of the roof were reproduced in the test.

In order to create the model, the natural frequencies and the corresponding modes of the roof were required, for which the relevant calculations had to be made. Ill. 3 shows the mode of a canopy moving backwards and forwards along the diagonal. Ill. 4 shows the mode of the diagonally opposite bay of the canopy moving up and down. When the roof was nearly completed, measurements of the vibration of the canopy in the south-west corner were carried out by the department for timber construction of the University of Technology, Munich. The measurements were made before the individual canopies were connected (see page 60).

The Fourier coefficient of the time series for the measured acceleration values shows the dominant frequencies. Since two sensors were fixed in different directions and at different points of the roof construction, a comparison of the two allows the appropriate mode to be attributed to the respective frequency. Ill. 1 shows a mean amplitude spectrum of the vertical acceleration. Ill. 2 shows that of the horizontal acceleration. In the spectrum of horizontal accelerations, the expected peak value at around 2.8 Hz does not occur. One may, therefore, conclude that there is a mode corresponding to this peak frequency, shown in ill. 1. The mode has no element of horizontal movement like the butterfly form in ill. 4.

The dimensions of the roof structure were based on calculations and wind-tunnel tests. Measurements made on the finished roof confirm the calculated characteristic dynamic values of the system.

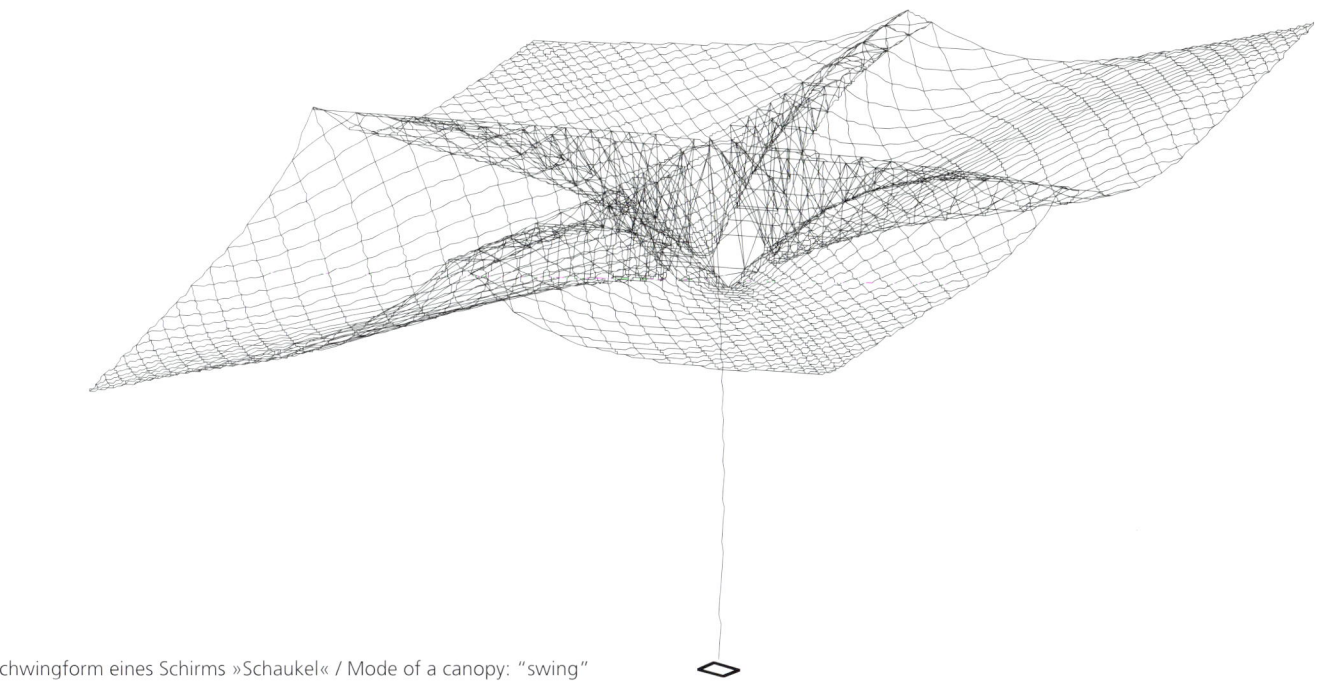

3 Eigenschwingform eines Schirms »Schaukel« / Mode of a canopy: "swing"

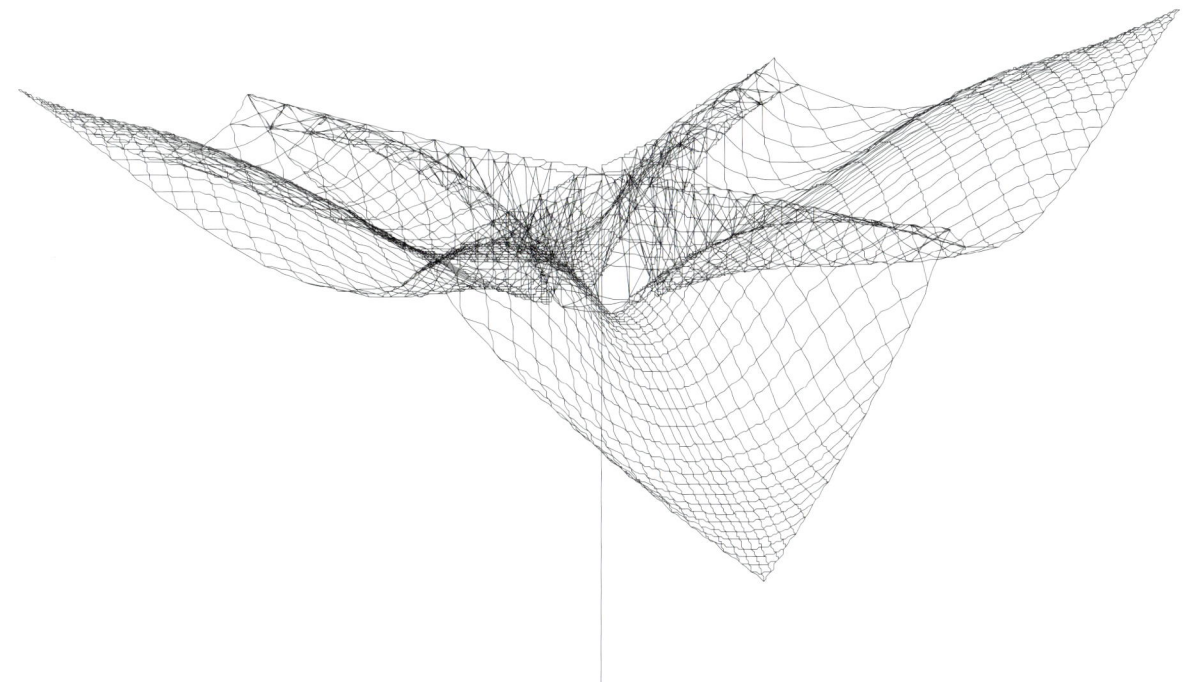

4 Eigenschwingform eines Schirms »Schmetterling« / Mode of a canopy: "butterfly"

Zeichnungen / Drawings: IEZ

Gitterschale vor der Montage / Lattice-grid shell prior to assembly

Versuche im Windkanal

Wind-Tunnel Tests

Jacques-André Hertig

Versuch im Windkanal:
Simulation einer Schneeverwehung am Dach.
Links während des Versuchs, rechts nach dem Versuch

Wind-tunnel test:
Simulation of snow drifts on roof.
Left: during the test; right: after the test

In den einschlägigen Regelwerken für Windlastannahmen sind keine Angaben zu Einwirkungen aus Windbeanspruchung bei weit auskragenden und doppelt gekrümmten Dächern zu finden. Insbesondere kann die Verteilung der aerodynamischen Kraftbeiwerte bei hintereinander stehenden bzw. unterschiedlich gruppierten Schirmen nicht bestimmt werden. Gleichermaßen gilt dies für auftretende Schneeumlagerungen.

Die Untersuchung des aerodynamischen Verhaltens der Schirmkonstruktion sowie der möglichen Schneeumlagerungen führte das Institut LASEN (Laboratoire de Systèmes Energetiques) der Eidgenössischen Technischen Hochschule in Lausanne durch.

Mit einem Grenzschicht-Windkanal wurden die verschiedenen Windspektren (unterschiedliche Geschwindigkeiten in verschiedenen Höhen und Böigkeit des Windes in Abhängigkeit von der Zeit) nachvollzogen. An maßstabsgetreuen Modellen im Windkanal wurde mit einem speziellen Meßverfahren, der sogenannten »Methode der Auflagerkräfte« (»Méthode des forces extrêmes«), die maximale Gesamtbeanspruchung auf eine Struktur durch Messung der Auflagerreaktionen bestimmt.

Die resultierenden aerodynamischen Kraftbeiwerte für die einzelnen Schalenflächen wurden über die statischen Gleichgewichtsbedingungen bestimmt und die Verformungen der Schalen mit einem Lasergerät gemessen.

Die Ergebnisse am Einzelschirm zeigten ein außergewöhnliches Verhalten. Die doppelt gekrümmte Form der Schalen wirkt aerodynamisch wie eine auf dem Kopf stehende Tragfläche eines Flugzeugs. Der dadurch auf der Schalenunterseite entstehende Unterdruck bewirkt eine zusätzliche Belastung der Schalenflächen nach unten.

Die Verformungen zweier diagonal gegenüber liegender Schalenflächen werden von den beiden anderen, ebenfalls diagonal gegenüberliegenden Schalenflächen direkt beeinflußt. Dadurch ergibt sich eine Art »Schmetterlingseffekt« (die gegenüberliegenden Schalen bewegen sich wie die Flügel eines Schmetterlings).

Der für die gesamte Struktur gültige dynamische Kraftbeiwert (c_{dyn}) wurde durch eine dynamische Analyse der gemessenen Spektren für die Verformungen und für die Auflagerkräfte ermittelt. Das Rechenmodell erlaubt auf das dynamische Verhalten der Schirme zu schließen und den statisch unbestimmten Schmetterlingseffekt auszuschließen.

Zur Simulation von Schneeverwehungen auf den Dachschalen wurden die Partikel, die den Schnee simulierten, gleichmäßig mit einer bestimmten Menge auf die Schirme verteilt. Durch Messung der nach der Windeinwirkung vorhandenen Mengen auf den einzelnen Schalen konnte die Zu- bzw. Abnahme der Schneemengen auf den einzelnen Dachschalen bestimmt werden. Damit wurden die möglichen ungleichmäßig verteilten Schneelasten aus windbedingten Schneeumlagerungen festgelegt.

Common standards that give the assumed values for wind loads contain no details of the effects of such loads on extensively cantilevered double-curved roofs. The distribution of the aerodynamic coefficients of forces for canopies laid out in series or in various groupings cannot be determined. The same applies to shifting snow loads.

The investigation of the aerodynamic behaviour of the present canopy structures and of potential shifts in snow-loading was conducted by the Institut LASEN (Laboratoire de Systèmes Energetiques) of the University of Technology in Lausanne.

The various wind spectrums (different speeds at different altitudes, and the gustiness of wind in relation to time) were reconstructed in a boundary-layer wind tunnel. The maximum overall loading on a structure was determined, using true-to-scale models in the wind tunnel and a special measuring technique: the so-called "extreme-load examination method" (*méthode des forces extrêmes*).

The resultant aerodynamic force coefficients for the individual shell surfaces were determined on the basis of conditions of structural equilibrium, and the deformation of the shells was measured with a laser probe device.

The results for a single canopy revealed exceptional behaviour. The double-curved form of the shells has an aerodynamic effect rather like that of the inverted wing of an aeroplane. The negative pressure thereby created on the underside of the shell results in additional downward forces on the shell areas.

The deformation in two diagonally opposite shell areas is directly influenced by the two other shell segments – also diagonally opposite each other. The result is a kind of "butterfly effect", in which the opposite shells move like the wings of a butterfly.

The relevant dynamic coefficient of forces for the entire structure (c_{dyn}) was calculated for the deformation and for the reaction forces by means of a dynamic analysis of the measured spectra. The calculation model permits conclusions to be drawn about the dynamic behaviour of the canopies and enables the statically indeterminate butterfly effect to be eliminated.

For the simulation of snow drifts on the roof shells, a certain amount of the particles used to simulate the snow were distributed evenly over the canopies. Measurements of the amounts present on the individual shells after the effects of wind allowed the increase or decrease to be ascertained. In this way, it was possible to determine a potentially uneven distribution of snow loading resulting from snow shifts caused by the wind.

Seite 53 oben Mitte: Blick von unten in die Schirme /
Page 53 top middle: View of canopies from below

Seite 53 oben rechts: Ausschnitt /
Page 53 top right: Detail

Innovationen im Holzbau

Innovations in Timber Construction

Norbert Burger, Julius Natterer

Anmerkungen zum konstruktiven Holzschutz

Holz weist im allgemeinen eine hohe Dauerhaftigkeit auf. Ohne chemischen Holzschutz kann sie allerdings nur bei Einhaltung konstruktiver Regeln des baulichen Holzschutzes, die die Besonderheiten des Baustoffes berücksichtigen, erreicht werden. Dies ist durch eine Vielzahl z. T. sehr alter Holzkonstruktionen dokumentiert. Beeinflußt wird die Dauerhaftigkeit auch durch die Wahl der Holzart.

Die Einhaltung der Regeln des konstruktiven Holzschutzes ist in jedem Holzbauprojekt oberstes Gebot. Da im Rahmen dieses Projektes auf einen chemischen Holzschutz vollständig verzichtet werden sollte, kommt dem konstruktiven Holzschutz besondere Bedeutung zu.

Aufgrund der enormen Abmessungen und der technologischen Eigenschaften wurde als Holzart für die Stützen Weißtanne gewählt. Mit der kompletten Überdeckung der Schirme und deren Bauteile durch eine Dachhaut ist ein vollständiger Regenschutz der Konstruktion gegeben. Die Dachhaut ist von allen Bauteilen abgehoben und »schwebt« in ca. 5 cm Abstand über den tragenden Bauteilen. Dadurch ist eine dauernde Umlüftung aller Bauteile gewährleistet. Nach vorübergehender Befeuchtung, beispielsweise durch Abtropfen von Tauwasser, können die Bauteile innerhalb kurzer Zeit wieder ganz austrocknen.

Ein kritischer Punkt ist die Fuge zwischen der Stahl-Grundplatte und der Hirnholzfläche am Stützenfuß; hier wurde das Hirnholz mit einem Latexanstrich versehen.

Windeinwirkungen sorgen für einen ständigen Luftaustausch zwischen den Bauteilen und damit für eine »aktive« Belüftung.

Brettstapelkonstruktionen

Die Konstruktion des EXPO-Daches bildet einen Höhepunkt der Entwicklung von Brettstapelkonstruktionen für den Holzschalenbau. Mit fast 28 m Länge in der Hauptdiagonalen vom Schalentiefpunkt an der Turmkonstruktion bis zur freien Ecke der Schale weist die Konstruktion die bisher größte realisierte Spannweite in dieser Bauweise auf.

Die Brettstapelbauweise aus horizontal vernagelten, verschraubten oder verleimten Brettern wird seit Ende der 60er Jahre bei ebenen und gekrümmten Flächentragwerken eingesetzt. Bretter, auch von geringerer Holzqualität, werden mit der flachen Seite aneinander gestellt bzw. gelegt und durch Vernageln, Verschrauben, Verdübeln oder Verleimen miteinander verbunden.

Statisch gesehen tritt bei Brettstapelkonstruktionen – ähnlich wie bei Brettschichtholz – ein Vergütungseffekt ein. Schwachstellen (z. B. Äste) nebeneinander liegender Bretter treten in der Regel nicht an der selben Stelle im Bauteil auf und sind somit besser verteilt. Ein dadurch entstehendes »soziales« Tragverhalten ermöglicht es, daß stärkere Bretter mit höherer Holzqualität einen Teil der Beanspruchungen der Bauteile von geringerer Holzqualität übernehmen.

Die Herstellung von Brettrippenschalen erfolgt schichtweise, Brettlage für Brettlage. Bei der Fertigung können die einzelnen Bretter um ihre Längsachse verdrillt und um ihre schwache Querschnittsachse gebogen werden. Dies ist beim Festlegen der Rippenverläufe wichtig.

Beim Herstellen der Rippenschale müssen die Rippenachsen sogenannten geodätischen Linien – der kürzesten auf der Schalenfläche liegenden Verbindung zweier Punkte der Schalenfläche – folgen. Mathematisch ausgedrückt ist dies der Fall, wenn die Hauptnormale der Linie in jedem Punkt (d. h. die Senkrechte zur Tangente in diesen Punkten) mit der Flächennormalen der Tangentialfläche in diesen Punkten parallel oder antiparallel ist.

Als Gesamtstruktur ergeben sich statisch hochgradig unbestimmte Systeme mit hohem Lastumlagerungspotential. Umlagerungsmöglichkeiten bestehen zwischen den einzelnen Brettlagen einer Rippe und auch zwischen benachbarten Rippen. Durch die sichtbaren Brettrippen und die damit sichtbare Struktur ist der Kräfteverlauf und die Lastabtragung im Tragwerk optisch greifbar.

Bei den Holzrippenschalen bestehen die Rippen aus gekrümmten Brettstapeln. Die einzelnen Brettlagen sind nachgiebig miteinander verbunden und laufen an den Rippenkreuzungspunkten wechselseitig durch. Die unterbrochenen Brettlagen sind an den Kreuzungspunkten jeweils als Füllbretter gestoßen. Die Aussteifung der Schale erfolgt durch eine oder zwei Schalungslagen, die aus diagonal zu den Rippen verlaufenden Brettern bestehen.

Die Schalenform ist während der Montagezeit entweder durch ausreichende punkt- oder linienförmige Abstützungen oder durch ein vollflächiges Lehrgerüst, wie bei den Schalen für das EXPO-Dach, vorgegeben.

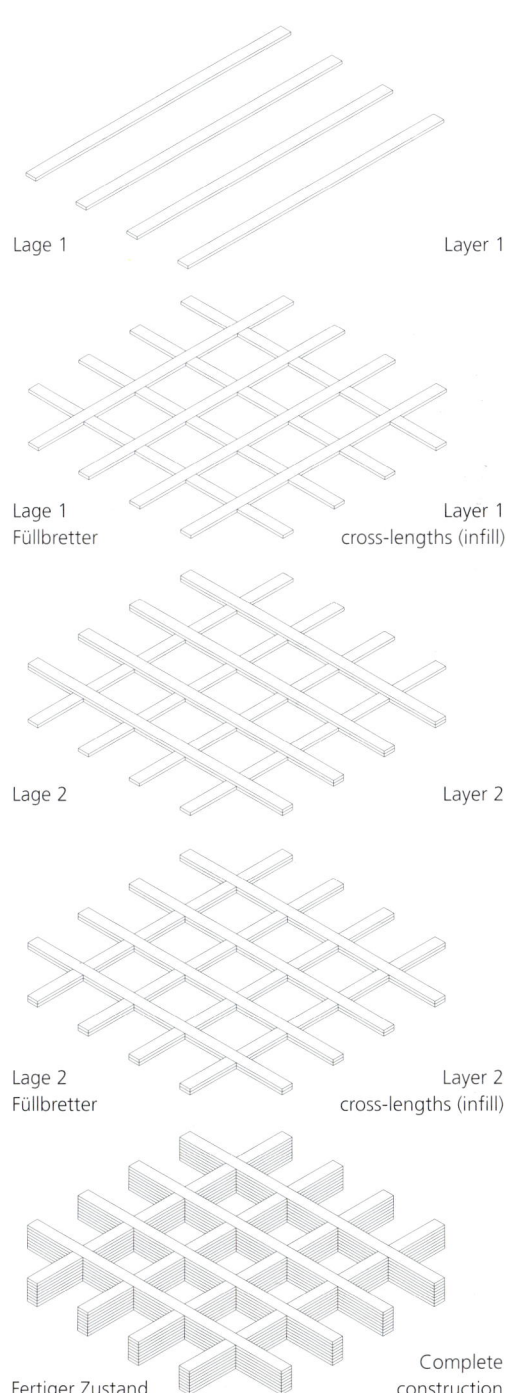

Brettstapelkonstruktion: Prinzip des Aufbaues / Stacked-plank construction: layering principle

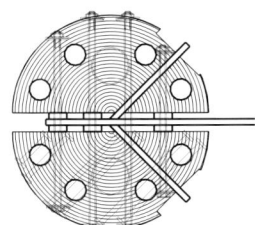

Anschluß Stützenfuß (+ 1,86 m) Dübel, Paßbolzen, Schwerter

Abutment at foot of column (+ 1.86 m): dowels, fitted bolts, fins

Verbindungsteile in verschiedenen Ebenen der Baumstämme in der Turmkonstruktion

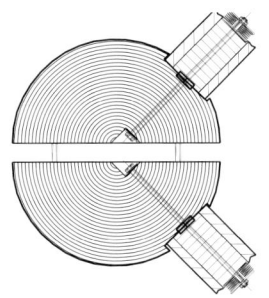

Anschluß untere Turmaussteifung (+ 5,58 m) Randhölzer mit Tellerfedern, Bolzen

Abutment of lower tower bracing members (+ 5.58 m): edge timbers with disc springs and bolts

Connectors in tree-stem columns at different levels of tower structure

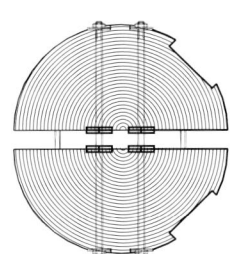

Regeldetail der Stammverbindung: Holzplatte mit Hartholzleiste als Abstandhalter, Bolzen, Einpreßdübel

Standard detail for connection of tree stems: timber board with hardwood strip as distance piece; bolts, dowels

Einsatz neuer, nicht durch Normen oder Zulassungen geregelter Bauweisen

Die Neuartigkeit der Konstruktion erforderte den Einsatz neuartiger Bauweisen, für die es keine allgemein gültigen Regelungen gibt oder für deren Einsatz bestehende Regeln in Form von Anwendungsnormen oder allgemeinen bauaufsichtlichen Zulassungen abgewandelt werden mußten.

Durch intensive Zusammenarbeit von Architekt, Tragwerksplaner, Prüfingenieur, Bauamt sowie ausführenden und zuliefernden Firmen konnten erforderliche Sonderkonstruktionen und -bauweisen ohne Zustimmung im Einzelfall ausgeführt werden.

Einbau der Rundholzstützen mit hoher Holzfeuchte
Gemäß DIN 1052 ist Holz maximal mit der später zu erwartenden Ausgleichsfeuchte (bei überdachten Holzbauteilen 15 ± 3%) einzubauen. Kann das Holz auch im eingebauten Zustand austrocknen, darf auch Holz mit höherer Feuchte verwendet werden, wenn bei den statischen Nachweisen die Festigkeitswerte um ein Drittel und die Steifigkeitswerte um ein Sechstel abgemindert werden.

Diese Regelung gilt im Grunde nur für höhere Holzfeuchten unterhalb des Fasersättigungspunktes. Die Stützen weisen hier Holzfeuchten deutlich oberhalb des Fasersättigungspunktes auf, so daß der übliche Erfahrungsbereich verlassen wurde. Auf die Austrocknungsmöglichkeiten und die entstehenden Zwangsbeanspruchungen im Bauwerk wurde daher beim Entwickeln der Konstruktion besonders geachtet.

Stützenanschlüsse in Halbrundholz mit hoher Holzfeuchte
Bei der Wahl der Verbindung für die Stützenanschlüsse spielte neben der Steifigkeit die Frage des Tragverhaltens bei hoher Holzfeuchte und dessen Veränderung bei zunehmender Austrocknung eine entscheidende Rolle. Der Anschluß wurde mit der Verbindung System Bertsche-BVD ausgeführt, für die eine allgemein bauaufsichtliche Zulassung betreffend die Anwendung in trockenem Holz existiert. Sie besteht aus einem profilierten Stahl-Ankerkörper, der in eine stirnseitige Sacklochfräsung eingebracht wird. Durch die Profilierungen verlaufen senkrecht zum Ankerkörper kreuzweise angeordnete Stabdübel. Der Ankerkörper hat in der Sacklochfräsung ausreichende Toleranzen zum Herstellen des Anschlusses. Nach Verfüllen des Hohlraums mit dem volumenneutralen Vergußmörtel aus dem System BVD wird der Zustand fixiert. Dadurch ergibt sich ein vollständiger Formschluß, so daß zum einen der bei nachgiebigen Verbindungen im Holzbau unvermeidliche Anfangsschlupf unterbunden wird und zum anderen die Verbindung eine sehr hohe Steifigkeit aufweist, die mit der Steifigkeit von Leimverbindungen vergleichbar ist.

Die Lastübertragung erfolgt vom Ankerkörper über die Stabdübel in den Holzquerschnitt und somit analog zu konventionellen Stabdübelverbindungen. Die Tragfähigkeit der Verbindung bei hoher Holzfeuchte ist daher nur von den Festigkeiten des Holzquerschnitts abhängig.

Mit zunehmender Austrocknung schwindet der Holzquerschnitt; es treten Risse auf. Ihr Einfluß auf die Tragfähigkeit sowie die Auswirkungen der eingebauten Stabdübel auf die Rißausbildung konnte durch Tragfähigkeitsversuche und vorhergehende Trocknung der Prüfkörper mit eingebautem Anschluß geklärt werden.

Brettrippen als mehrteilige, nachgiebig verbundene Querschnitte
Für mehrteilige, nachgiebig verbundene Querschnitte der Brettrippen existieren keine geregelten Berechnungs- und Nachweisformeln. Diese wurden mit Hilfe vorhandener Literatur für den vorliegenden Fall mit acht bis zehn Teilquerschnitten entwickelt.

Anschluß Querrippen an Längsrippen im Kehlbereich als Nebenträgeranschluß mit gekreuzten Schraubenpaaren
Um eine ausreichende Tragfähigkeit der Längsrippen im Kehlbereich der Rippenschale, in dem sich ein wesentlicher Teil der Kräfte aus der Schale konzentriert, zu erhalten, dürfen die einzelnen Brettlagen nicht unterbrochen werden. Dies bedeutet, daß hier alle Brettlagen der Längsrippen durchlaufen müssen und nicht von Brettlagen der Querrippen gekreuzt werden dürfen. Neben der ausreichenden Tragfähigkeit stand hier die möglichst einfache Herstellbarkeit im Vordergrund.

Der Anschluß der Querrippen an die Längsrippen im Kehlbereich der Schale wurde in Abwandlung einer bestehenden Zulassung von kreuzweise und im 45°-Winkel angeordneten Sonderschrauben für Kraftübertragungen rechtwinklig zur Faserrichtung des Holzes ausgeführt. Bei der vorliegenden Anwendung verlaufen die Schraubenpaare durch das Hirnholz der Querrippen. Als Grundlage für das Bemessen war die Auszugsfestigkeit der Schrauben aus Hirnholz- und Seitenflächen bei unterschiedlichen Einschraubwinkeln maßgebend.

Laschenstöße mit Anschlußblechen und eingeschossenen Sondernägeln
Die Bretter der Schalungslagen laufen nicht in voller Länge durch, sondern sind bis zu zweimal gestoßen. Die Beanspruchungen in den Schalungslagen sind so groß, daß eine Überleitung der Kräfte an der Stoßstelle auf die benachbarten Schalungsbretter nicht möglich ist. Jeder Stoß mußte daher für die volle Kraft ausgebildet werden.

Für die obere Schalungslage konnten dazu Holzlaschen verwendet werden. Ausgeführt wurden die Laschen wegen der geringeren erforderlichen Laschendicke in Bau-Furnier-Sperrholz (BFU). Für die untere Schalungslage kamen als Stoßlaschen jedoch nur dünne Stahlbleche in Frage.

Durch die Stahlbleche sollten wegen des einfacheren und schnelleren Herstellens der Verbindung die Nägel in das Holz »eingeschossen« werden.

Die Ausführbarkeit als eingeschossene Nagelverbindung wurde in mehreren Versuchen getestet. Dabei wurde der Druck der Nagelpistole variiert und schließlich so definiert, daß die Nägel vollständig eingeschossen werden, die Gefahr eines Aufspaltens der Bretter trotz einer hohen Nagelanzahl aber gering ist.

Mechanische Verbindung der Brettschichtholz-Deckel mit den Gurtquerschnitten durch Schubverbinder
Auf Wunsch der ausführenden Firmen sollten im Kastenbereich des Kragträgers die horizontalen Querschnittsteile (»Deckel«) mit den Gurtquerschnitten durch mechanische Verbindungen angeschlossen werden. Der Anschluß erfolgte mit dem Verbindungssystem Bertsche-VA. Es besteht aus Stahl-Schubverbindern, die in Ausfräsungen liegen und die Schubkräfte zwischen den Querschnittsteilen übertragen. Die Ausfräsungen sind etwas größer als die Schubverbinder, so daß ausreichende Toleranzen zum Herstellen der Verbindung vorgesehen werden können. Die Hohlräume sind mit Vergußmörtel BVD ausgegossen. Diese Verbindung weist daher eine sehr hohe Verbundsteifigkeit auf.

Verleimungen
Für folgende Bauteile bzw. Baugruppen waren Verleimungen erforderlich:
- 7 Rippen in der Hauptdiagonale der Schale auf ganzer Länge (Schraubpreßverleimung lagenweise auf dem Lehrgerüst)
- Rippenanschlüsse bis zum ersten Rippenkreuzungspunkt (Schraubpreßverleimung lagenweise auf dem Lehrgerüst)
- Rippen in höher beanspruchten Teilbereichen (Schraubpreßverleimung lagenweise auf dem Lehrgerüst)
- Randträger der Schale (I-Querschnitt durch Blockverleimung von gekrümmten Teilquerschnitten aus Brettschichtholz)
- doppelt gekrümmte BFU-Schale im Kehlbereich (vollflächige Schraubpreßverleimung von 8 mm dicken BFU-Platten zu einer doppelt gekrümmten Schale, Gesamtdicke 48 mm)
- Obergurt des Kragträgers (üblicher gerader Brettschichtholzquerschnitt)
- Untergurt des Kragträgers (besonderer Verlauf der Brettlamellen, z. T. blockverleimt)
- »Deckel« zwischen den Kragträgergurten im geschlossenen Bereich (übliche gerade Brettschichtholzquerschnitte)
- Verleimung von Ober- und Untergurt im äußeren Bereich des Kragträgers (Preßdruck über durchgesteckte Gewindestangen und außen liegende Stahlplatten)

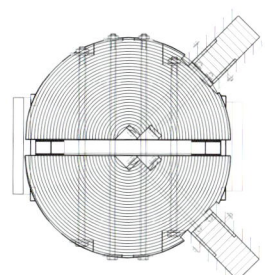

Anschluß der aussteifenden Schichtholzplatten und der Querzugsicherung (+ 7,99 m) Paßbolzen

Abutment of laminated glued timber sheet bracing and transverse tension stays (+7.99 m) with fitted bolts

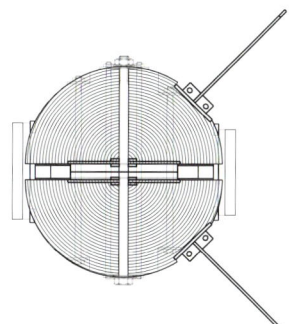

Mittlerer Knotenpunkt (+ 12,15 m): Schichtholz-Aussteifung, Querzugsicherung, Paßbolzen, Stahl-Sonderteil

Middle node (+12.15 m): laminated sheet bracing, transverse tension stays, fitted bolts, special steel component

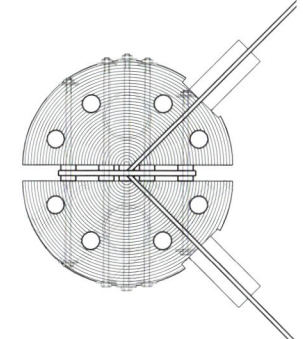

Anschluß am Kopfpunkt (+ 17,20 m) Paßbolzen, Dübel

Abutment at column head (+ 17.20 m) fitted bolts, dowels

- Vollwandstege und Schotten des Kragträgers (Schraubpreßverleimung von zwei Lagen Furnierschichtholz)
- Querschnitte des K-Verbands im Kragträger (übliche gerade Brettschichtholzquerschnitte)
- Aussteifungsscheiben der Turmkonstruktion (Schraubpreßverleimung von Furnierschichtholzscheiben mit üblichen geraden Brettschichtholzquerschnitten).

Schraubpreßverleimungen können analog zu den in DIN 1052 geregelten Nagelpreßleimungen ausgeführt werden. Schrauben dürfen im Gegensatz zu Nägeln (Ausnahme: Sondernägel) planmäßig auf Zug beansprucht werden und haben ein besseres Anpreßverhalten, so daß der erforderliche Anpreßdruck mit einer gegenüber der Nagelpreßleimung geringeren Anzahl von Schrauben erreicht werden konnte. Die Schrauben sind im Raster von 10 cm x 10 cm eingeschraubt (vgl. dazu Nagelpreßleimung nach DIN 1052: 8 cm x 8 cm).

Um ein vollständiges und ungestörtes Aushärten des Leims zu gewährleisten, ist eine Umgebungstemperatur von 20° C einzuhalten; die Bauteile dürfen während der ersten 12 Stunden keinen Erschütterungen ausgesetzt werden. Da zum Herstellen der Schalen vollflächige Lehrgerüste in einer Messehalle aufgebaut waren, konnte die Einhaltung dieser Anforderungen auch für das Verleimen der Rippenquerschnitte gewährleistet werden.

Notes on Constructional Means of Timber Protection

Timber possesses great natural durability. Without chemical preservatives, however, its durability can be exploited to the full only if constructional rules are observed that take account of the special characteristics of the material. This is confirmed by a large number of timber structures, some of which are extremely old. Durability is also influenced by the type of wood chosen.

Observing the rules of constructional timber protection is of paramount importance for any structure in which this material is used. Since all forms of chemical preservation were to be avoided in the present scheme, the rules of timber protection by constructional means assumed a special importance.

In view of the enormous dimensions involved, and the tehnological properties the timber used for the columns was to be taken only from silver fir trees. Complete protection against rain was ensured by covering the entire area of the canopies and their constructional components with a waterproof skin. The roof skin is raised over the entire structure by approximately 5 cm and "floats" above the load-bearing elements. This also ensures that air can circulate around all building components. Should any members be exposed to temporary moisture – through the dripping of condensation, for example – they can dry out completely within a very short time. The junction between the steel base plate and the end-grain surfaces of the timber
represents a critical point, but this was overcome by applying a coat of latex.

Wind action ensures a constant exchange of air and thus an "active" ventilation of the structure.

Stacked-Plank Construction

The EXPO roof marks a high point in the development of stacked-plank construction for timber shell structures. With a length of almost 28 m along the main diagonal, from the lowest point of the shell next to the tower to the freely cantilevered corners, the structure has by far the greatest span ever realized in this form of construction.

Stacked-plank construction consists of horizontally nailed, screwed or glued boards and has been in use since the end of the 1960s for flat and curved planar load-bearing structures. Lower grades of timber can be used for the planks, which are stacked face to face or laid on top of each other and joined together by nailing, screwing, dowelling or gluing.

Structurally, there is a bonus effect with stacked-plank construction similar to that achieved with glued-laminated timber. Weaknesses such as knots do not normally occur at the same point in adjacent boards within an element, but are likely to be more evenly distributed. This results in a "social" quality in the load-bearing behaviour of the members: the stronger planks of higher-quality timber assume part of the loading from the lower-quality planks.

Shells with stacked-plank ribs are constructed layer by layer, plank by plank. During assembly, the individual planks can be twisted about their longitudinal axis and bent about their weak cross-sectional axis.

In timber ribbed shells, the ribs consist of curved stacks of planks. The individual layers are compliantly connected to each other and continue across the points of intersection between the ribs in alternate directions. The alternate layers that are cut off at these points are butt jointed at the sides. The shell is braced by one or two layers of boarding fixed diagonally to the rib lattice.

During the assembly period, the shell forms have to be maintained by providing adequate point or linear supports or by means of centring over the entire area, as was the case with the shells of the EXPO roof.

The rib axes of lattice-ribbed shells have to follow what are known as geodesic lines – the shortest route between two points across the surface of the shell. In mathematical terms, this occurs when the principal normal of the line at every point (i.e. perpendicular to the tangent at these points) is either parallel or antiparallel to the surface normals of the tangential plane at these points.

Viewed in its entirety, the structure reveals statically extremely indeterminate systems with a high load-redistribution potential. Scope for the redistribution of loads exists between the individual rib planks as well as between adjacent ribs. By allowing the plank ribs and the structure they form to remain visible, the system of forces and the transmission of loads within the structure is made legible.

Querschnitte der Halbstämme mit Einfräsungen und Verbindungsteilen

Cross-sections through semicircular logs with sinkings and connecting elements

Versuchsanordnung: Zugbeanspruchung von Verbindungsteilen System BVD an Weißtannen-Halbstämmen

Testing layout: tension loading of BVD system connecting elements in silver fir half-stems

Use of Innovative, Non-Regulated Types of Construction

The innovative nature of the roof structure necessitated the use of new types of construction for which no generally valid regulations exist or for which existing rules, standards or methods approved by building supervisory authorities had to be modified.

A close collaboration between the architects, structural engineers, proof engineers, building authorities, the firms responsible for the execution of the work and the suppliers made it possible to realize the necessary special structures and forms of construction without having to seek approval in specific cases.

Assembly of round timber columns with high moisture content

In accordance with German Standards (DIN 1052), timber may be built into structures only with its ultimate equilibrium moisture content (in the case of covered elements in timber construction: 15 ± 3 per cent). If the timber can continue to dry out after it has been incorporated into the structure, members with a higher moisture content may be used, in which case, however, the structural calculations of the strength values have to be reduced by a third and the stiffness values by a sixth.

This rule applies in principle only to higher timber moisture levels below the fibre-saturation point. The columns used in this scheme, however, have a moisture content well above the fibre-saturation point and were not covered by the range of values applicable in standard practice. In developing the structure, therefore, special attention was paid to the scope for drying the timber and the resultant constraining forces.

Connections of columns made up of semicircular stems with high moisture content

Two factors were of importance in determining the form of the column connections at the head and foot: the rigidity; and the load-bearing behaviour of members with a high moisture content, as well as changes in this behaviour as a result of the continuing process of drying out. Connections were made with the Bertsche BVD system, general permission for the use of which exists in conjunction with dry, seasoned timber. The system consists of a moulded cylindrical steel anchor fixed in a pocket hole cut in the end grain of the timber. Dowel-type fasteners are inserted at right angles to each other through the anchor member. The pocket holes allow adequate space to make the connections with the anchor. After the void has been filled with a grouting mortar not subject to autogenous volume change – also from the BVD system – the anchor is fixed firmly in position. The result is a rigid connection, which eliminates the initial slipping that is usually unavoidable with compliant timber connections. The connections possess great rigidity, comparable to that of adhesive fixings.

Loads are transmitted from the anchor member via the dowel-type fasteners into the timber cross-section in an analogous form to conventional dowel-type connections. The load-bearing capacity of such connections in timber with a high moisture content is, therefore, dependent solely on the strength of the timber cross-section.

With increasing drying out of the timber, the cross-section shrinks, and cracks occur. The influence of these cracks on the load-bearing capacity and the effects of the steel dowels on the development of the cracks were clarified by tests of the bearing strength and by the prior drying of the test volumes with inbuilt anchor elements.

Stacked-plank ribs as multiple, compliantly connected cross-sections

No standard formulae for calculations or proof of stability exist for the multiple, compliantly connected cross-sections of the ribs in stacked-plank construction. Based on existing literature, the requisite formulae were developed specially for the present scheme, in which eight- to ten-layer cross-sections occur.

Connections of cross-ribs to longitudinal ribs in the throat area of the shell in the form of secondary beam connections with crossed pairs of screws

To achieve adequate structural strength in the longitudinal ribs in the throat area of the lattice shell, where a major part of the loads in the shell are concentrated, it was important to ensure that the plank layers were not interrupted. In other words, in this area, all board layers of the longitudinal ribs had to run through continuously and were not to be intersected by the planks of the cross-ribs. In addition to ensuring adequate load-bearing capacity, the form of assembly was to be as simple as possible.

The connections of the cross-ribs to the longitudinal ribs in the throat area of the shell were executed – as a variation on an existing approved type of construction – with special screws inserted at an angle of 45° and crossing each other for load transmission at right angles to the fibre direction of the wood. In the present case, the pairs of screws are driven into the end grain of the cross-ribs. The basis of measurement was the pull-out resistance of the screws – set at various angles – from the end-grain wood and the side faces of the members.

Butt joints with fishplate connectors and special shot-in nails

The planks in the top layers of boarding do not extend over the entire breadth of the roof shell, but have up to two butt joints in their length. The stresses in these boarding layers are so great that a transmission of loads at the points of abutment with the adjoining boards is not possible. Every butt joint, therefore, had to be designed to bear the full loading.

Wood fishplates were used for this purpose on the upper layer of the roof boarding. In view of the slender dimensions required, laminated construction board was used for this purpose. Thin steel sheets were the only possible form of construction for the fishplates over the lower layer of boarding.

To permit a simpler, quicker form of connection, the nails were to be "shot" into the timber through the steel sheets. The feasibility of executing shot-nailed connections of this kind was tested in a number of trials, in the course of which, the pressure of the nail gun was varied. The pressure was finally fixed at a level at which the nail is shot in its entirety into the timber. Despite the great quantity of nails used, the danger of splitting the boards is relatively small.

Mechanical connection of the laminated timber "lid" to the chord cross-sections with shear connectors

At the wish of the firms executing the work, the horizontal elements or "lids" of the box sections of the cantilevered trusses were connected to the chord cross-sections with mechanical fixings. The connection was made with the Bertsche-VA connecting system, which consists of steel shear connectors fitted into pockets cut in the timber members. The connectors transmit the shear forces between the different members. The pockets were cut somewhat larger than the shear connectors to allow adequate space to make the connections. The voids were subsequently grouted with BVD mortar. This form of connection has a very high bonding stiffness.

Glue bonding

Glued joints were necessary for the following building components or elements:
- seven ribs along the main diagonal of the shell over the entire length (screw-fixed and glued connections layer by layer on centring)
- rib connections up to the first rib intersection (screw-fixed and glued connections layer by layer on centring)
- ribs in areas subject to heavier loading (screw-fixed and glued connections layer by layer on centring)
- edge beams of the shell (I-section: block gluing of curved parts of the cross-sections in laminated timber construction)
- double-curved laminated construction-board shell in the throat area (screwed and glued bonding over full area with 8 mm laminated construction board to form a double-curved shell with 48 mm overall thickness)
- upper chords of cantilevered trusses (normal straight laminated timber sections)
- lower chords of cantilevered trusses (special line of laminated layers; partly block glued)

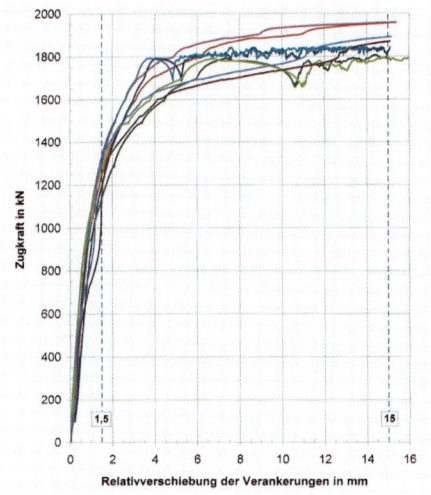

Randträger der Kragarme mit Ausfräsungen für die Verbindungsteile
edge beams of cantilevered members with pockets cut for connections

◁ Diagramm: Zugkraft-Auszugverhalten der Verankerungen System BVD bis zur Höchstlast
◁ Graph of tension loading (pull-out behaviour) of BVD anchor system up to failure point

- cover between chords of cantilevered trusses over closed length (normal straight laminated timber sections)
- upper and lower chords of cantilevered trusses over outer lengths (fixed with threaded rods through timber members and with outer steel plates)
- solid webs and lateral bracing of cantilevered trusses (two layers of laminated wood sheeting screw-fixed and glued)
- K-bracing members in cantilevered trusses (normal straight laminated timber sections)
- bracing slabs to tower (laminated timber sheeting screwed and glued to normal laminated timber sections)

Screw-fixed and glued connections can be executed in a similar way to nailed and glued connections, as recognized by German Standards (DIN 1052). In contrast to nails (with the exception of special nails), it is permissible to subject screws to controlled tension stresses. They also have a better bonding action, so that the requisite pressure can be achieved with a smaller number of screws than with the nails that would be necessary in a nailed and glued connection. The screws are fixed at grid spacings of 10 x 10 cm; (cf. nailed and glued connections in compliance with DIN 1052 with nails at 8 x 8 cm centres).

To ensure a complete, undisturbed setting of the glue, an ambient temperature of 20 °C has to be maintained. The building components should not be subject to any impact shocks or vibration during the first 12 hours after gluing. Compliance with these requirements for gluing the members of the structure, including the ribs, was ensured by assembling the shells on full-area centring in one of the trade-fair halls.

Experimentelle Untersuchungen zur Tragfähigkeit der Anschlüsse System BVD für die Zugverankerung der Stützen

Werner Kelletshofer, Robert Spengler

Prüfkörper, Ankerkörper

Zur Durchführung der erforderlichen Tragfähigkeitsversuche wurden dem Materialprüfungsamt für das Bauwesen der TU München 4,0 m lange Halbstämme aus Tannenholz zur Verfügung gestellt (durch den Kern längsgetrennte Vollstämme, Durchmesser etwa 85 cm). An beiden Enden dieser Halbstämme waren jeweils vier sogenannte BVD-Ankerkörper eingebaut. Diese sind wesentliche Teile des sogenannten Verbindungssystems BVD, für welches seit 1998 eine Allgemeine Bauaufsichtliche Zulassung besteht. Da diese Zulassung in strenger Auslegung für die vorgesehene und sehr spezielle Verankerungssituation (relativ feuchtes Holzmaterial bei der Herstellung, gemeinsame Tragwirkung von vier BVD-Ankerkörpern pro Halbstammende, zwei Ankerkörper überdeckende Stabdübelreihen, Halbkreisquerschnitt mit unterschiedlichen Flächenanteilen pro Anker u. a.) nicht zweifelsfrei anwendbar ist, waren entsprechende experimentelle Überprüfungen geboten.

Herstellen der Prüfkörper

Um eventuelle Einflüsse aus der Holzfeuchte auf das Schwind- bzw. Tragverhalten der BVD-Verbindung beurteilen zu können, wurde bei den Halbstämmen zum Zeitpunkt des Einbaues der Ankerkörper die jeweilige Holzfeuchte an verschiedenen Stellen gemessen. Hierbei wurden Holzfeuchtewerte von etwa 35 ± 5% festgestellt.

Im Zuge der anschließenden Kammertrocknung sowie der Zwischenlagerung im Prüfhallenklima (etwa 20° C, 50% relative Luftfeuchtigkeit) bis zum Zeitpunkt der Tragfähigkeitsversuche hat sich in den Prüfkörpern eine oberflächennahe Holzfeuchte von etwa 20 ± 5% eingestellt (Meßtiefe ca. 30 mm).

Prüf- und Meßvorrichtung

Die horizontal angeordnete Prüfvorrichtung ermöglicht die Einleitung einer Zugkraft von maximal 2000 kN, welche über ein festes sowie ein bewegliches Widerlager abgetragen werden mußte.

Von wesentlicher Bedeutung war das Last-Auszug-Verhalten der jeweils gemeinsam tragenden 4 BVD-Ankerkörper. Diesbezüglich waren (außer den Meßwertgebern zum Registrieren der Prüfkraft) an jeder Verankerungsstelle Meßwertgeber zur Messung der Auszugwege angeordnet.

Versuchsdurchführungen

Jeder der Prüfkörper wurde zunächst stetig bis auf ein Lastniveau von 400 kN und dann (nach Entlastung auf 100 kN) durch eine anschließende sechsmalige Wechselbeanspruchung zwischen 200 kN und 400 kN belastet. Nach vollkommener Entlastung folgte der eigentliche »Bruchversuch«, durch stetiges Belasten bis zum Versagen des Verbindungssystems BVD bzw. bis zum Überschreiten einer vorgegebenen Auszuggrenze (Prüfnorm) von 15 mm.

Beurteilung des Tragverhaltens

Im Rahmen der »Bruchversuche« haben sich erst ab etwa 1000 kN Zugkraft mehr oder weniger deutliche Schädigungen eingestellt, welche sowohl an den »einsinkenden« Enden der Stabdübel als auch an auftretenden »Spaltrissen« wahrgenommen werden konnten. Außerdem haben nun die Auszugwege der BVD-Ankerkörper bis zur Höchstlast überproportional zugenommen.

Die Ergebnisse zeigen aber, daß sich die Auszugwege (bis weit über das für dieses Verbindungssystem anzusetzende zulässige Lastniveau hinaus) annähernd proportional zur Last einstellen und bis zum Gebrauchslastniveau (400…600 kN) 0,5 mm nicht überschritten haben.

Einen offensichtlich günstigen Einfluß auf das Tragverhalten hatte die kreuzweise und teilweise zwei Ankerkörper übergreifende Stabdübelanordnung (Gitterstruktur), nicht nur im Hinblick auf die erreichten Höchstlasten (1790…1865…1960 kN), sondern auch hinsichtlich des Risikos der vollständigen Aufspaltung des Verankerungsbereiches.

Das Verbindungssystem BVD in der hier geprüften Ausführung kann als eine fast dehnstarre Verankerungsweise (1160…1340…1880 kN pro mm) und hinsichtlich der Versagensart sowie des Verformungsverhaltens auf Höchstlastniveau als ein Verbindungssystem mit sehr »gutmütigen« Trageigenschaften eingestuft werden.

Experimental investigations of the load-bearing capacity of the BVD system connections for the tension anchoring of the columns

Test samples and anchor elements

In order to carry out the necessary load-bearing trials, the Materials Testing Office for Building Construction at the University of Technology, Munich, was provided with 4-metre-long semicircular sections of fir tree stems (full fir trunks with a diameter of roughly 85 cm cut in half lengthways through the core). Four BVD anchor elements were inserted in both ends of the halved stems. The elements are important components in the BVD connection system, general permission for the use of which was granted by building supervisory authorities in 1998. Experimental trials were necessary, since, strictly speaking, the permission granted for the use of this system does not apply to the special fixing situation encountered in this scheme (timber with a relatively high moisture content at the construction stage; shared load-bearing function with four BVD anchor elements at each end of the halved stems; rows of dowels over two anchor elements; semicircular cross-section with different unit areas for each anchor, etc.). The system cannot be applied without certain reservations, therefore.

Preparing the testing samples

In order to assess the possible influence of high moisture content in the timber on the shrinkage and on the load-bearing behaviour of the BVD connections, the moisture levels were measured at various points in the halved stems at the time when the anchor elements were

Vorfertigung des Kragträgers im Werk /
Prefabrication of cantilevered truss at works

Arbeitsvorbereitung mit weiter entwickelter CAD/CAM-Software

Advanced CAD/CAM Software Applications

Martin Pfundt

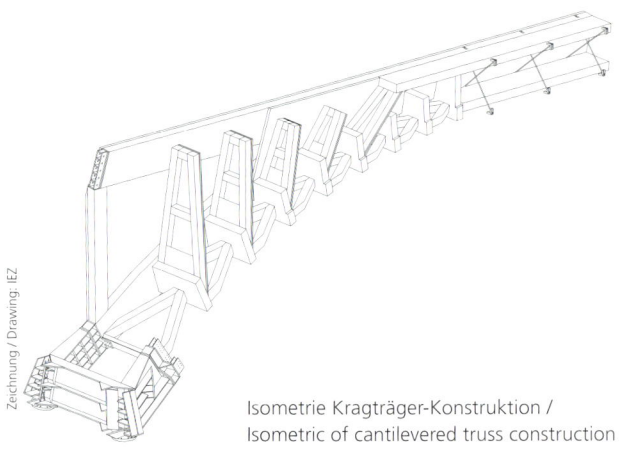

Isometrie Kragträger-Konstruktion /
Isometric of cantilevered truss construction

inserted. Moisture levels of roughly 35 per cent (± 5 per cent) were measured. In the course of the ensuing kiln drying and the storage period in the testing hall environment (approx. 20 °C at 50 per cent humidity) up to the time when the load-bearing trials were carried out, a moisture content of roughly 20 per cent (± 5 per cent) was measured in the surface layers of the trial samples (measurement depth approx. 30 mm).

Testing and measurement equipment
The testing equipment, laid out horizontally, allows a maximum tension loading of 2,000 kN to be applied, which was to be transmitted to a fixed and a flexible abutment.

Of great importance was the load-deflection behaviour of the four BVD anchor elements, which bear the loads jointly. In addition to the devices for measuring the testing loads, therefore, gauges for the measurement of the displacement were attached to every fixing point.

Testing process
Each of the test samples was initially subjected to a constant loading of 400 kN and then, after reducing the load to 100 kN, restressed six times with loads varying between 200 kN and 400 kN. After the samples had been relieved of all loading, they were finally subjected to the actual "breaking test", carried out by applying constant loading up to the point of failure of the BVD anchor system or to the point where a predetermined deflection limit (testing norm) of 15 mm was exceeded.

Assessment of load-bearing behaviour
In the context of the tests to failure, clear damage became evident only above roughly 1,000 kN tension loading. This became apparent through the ends of the dowel-type fixings sinking in and through the appearance of splits. In addition, the displacement of the BVD anchor elements increased disproportionately up to the maximum loading.

The results show, however, that the deflection remained roughly proportional to the loading applied (to a point far in excess of the permissible loading for this connection system) and did not exceed 0.5 mm up to the level of the working load (400...600 kN).

The cruciform arrangement of the dowels, which extend over two anchor elements at certain points (grid-like layout), evidently had a positive effect on the load-bearing behaviour, not just in respect of the maximum loads achieved (1,790...1,865...1,960 kN), but also in terms of the risk of a complete splitting of the members in the area of the anchorings.

The BVD connecting system in the form tested here can be classified as a virtually rigid means of anchoring components (1,160...1,340...1,880 kN per mm); and in view of the nature of the failure and the deformation behaviour at the maximum loading level, it may be regarded as a connecting system with extremely "benign" load-bearing properties.

Mit dem 3D-CAD/CAM-System »cadwork« ist es möglich, beliebig komplexe Geometrien zu erzeugen und zu bearbeiten. Die generierten Bauteile können automatisch in Listen für Materialdisposition oder Fertigung ausgegeben werden. Darüber hinaus sind die notwendigen Werkstattzeichnungen oder Daten zum Ansteuern gängiger Abbundanlagen einfach und übersichtlich zu erzeugen.

Zur Modellierung der Rippenschale
Grundlage für die Bearbeitung dieses Projektes war eine von den Tragwerksplanern erzeugte Datei mit den x, y, z-Koordinaten der Schnittpunkte der Rippen des statischen Tragwerksmodells der Schalenkonstruktion. Der Import dieser Datei in »cadwork« lieferte die Position der Kreuzungspunkte der Schalenrippen untereinander. Auf der Grundlage dieser Geometrievorgabe wurden die Rippen konstruiert, die wegen ihrer Fertigung aus durchgehenden Brettlamellen nur um ihre Längsachse verdreht und einfach gekrümmt sein dürfen. Sie wurden in »cadwork« als Spline-Körper modelliert.

Zur Modellierung der Randträger
Die Randträger der Rippenschale sind ebene, parabelförmige Bauteile mit einem I-Querschnitt von 600 x 640 mm² über die gesamte Länge von ca. 19 m; sie wurden aus 5 Teilen zusammengeleimt. An beiden Enden befinden sich geschweißte Stahlteile, welche die Randträger untereinander und mit den Kragträgern verbinden. An der Innenseite zwischen den Flanschen befinden sich Stahlteile, an denen die Rippen angeschlossen sind.

Die Schalenränder am Kragträger, die mit den gekrümmten Untergurten der Kragträger verbunden sind, sind ebenfalls eben und parabelförmig und haben bei einer Breite von 220 mm Querschnittshöhen zwischen 680 und 1020 mm. Die Oberkante der Randträger liegt in der gekrümmten Ebene der Schalenrippen, so daß die auf den Rippen befestigte Brettschalung flächig aufgelegt werden

konnte. Die Neigung variiert zwischen ca. 45° an der Kragarmspitze und ca. 64° am Kehlstahlteil. Die in »cadwork« aus der Schalengeometrie entwickelten Trägergeometrien wurden über die DXF-Schnittstelle exportiert und auf einer 5-Achsen-CNC-Anlage (Computerized Numerical Control) gefertigt. Die Produktionsgenauigkeit betrug ± 1 mm.

Zur Modellierung des Lehrgerüstes
Die Lehrgerüstrippen sind parabelförmig gekrümmt und haben durch ihre Breite von 80 mm zudem eine parabelförmige Abgratung, die ihre Neigung ständig verändert. Diese Lehrgerüstbauteile, die am Ende nirgendwo sichtbar sind, mußten für die Fertigung ebenso wie die Schalenbauteile bis auf ± 2 mm genau sein.

Die für die Verbindung der Bauteile untereinander eingesetzte Vergußtechnik stellte besondere Anforderungen an die Arbeitsvorbereitung. Die für den vollständigen Verguß notwendigen Befüll- und Entlüftungsbohrungen führten dazu, daß im eingebauten Zustand identische Bauteile, bedingt durch die unterschiedliche Anordnung der Bohrungen, völlig unterschiedlich gefertigt werden mußten.

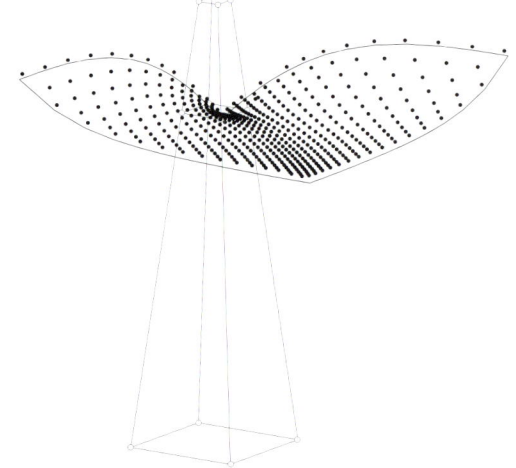

Geometrische Anordnung der Knotenpunkte der Gitterschale

Geometric layout of lattice shell nodes

Kragträger im Werk

Cantilevered truss at works

▷ Anschluß Gitterschale an Kragträger

▷ Abutment between lattice shell and cantilevered truss

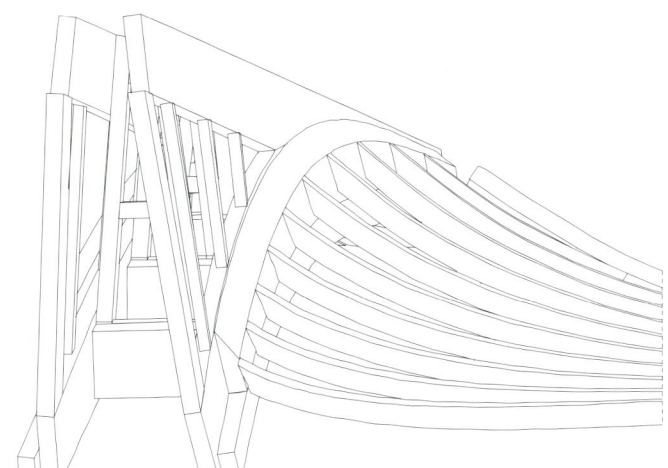

Zur Arbeitsvorbereitung gehörte ebenfalls die Simulation der verschiedenen Einbausituationen. Dabei mußte darauf geachtet werden, daß die Reihenfolge des Zusammenbaues nicht zu Kollisionen mit anderen Bauteilen führte und bestimmte Höhen nicht überschritten wurden, da sonst unter Umständen der Kran nicht mehr eingesetzt werden konnte.

Using the 3-D CAD/CAM system cadwork, it was possible to generate and develop geometries of any complexity. Lists of building elements designed in this way can be automatically issued for ordering materials and for the manufacturing process. In addition, the requisite workshop drawings or data can be produced for the operation of standard cutting equipment.

Modelling the ribbed shells

The basis of the work in this phase of the scheme was a file produced by the structural engineers, containing the x, y and z co-ordinates for the points of intersection of the ribs of the structural model of the canopy shell construction. The import of this file into *cadwork,* allowed the positions of the points of intersection between the various shell ribs to be determined. The ribs were constructed on the basis of this geometric model. In view of the fact that they are made up of continuous layers of boarding, they may be twisted only about their longitudinal axis and curved only in a single direction. They were modelled as "spline volumes" in cadwork.

Modelling the edge beams

The edge beams of the ribbed shells are planar, parabolic elements with an I-shaped cross-section 600 mm wide and 640 mm deep over their entire length (approximately 19 metres). They are assembled from five glued sections. At both ends are welded steel elements that connect the edge beams to each other and to the cantilevered trusses. On the inner face, between the flanges, are steel fixing elements for the shell ribs.

The edges of the shells connected to the curved lower chords of the cantilevered trusses are also planar and describe a parabolic curve. They have a width of 220 mm and a cross-sectional depth of between 680 and 1020 mm. The tops of the edge beams lie in the curved plane of the shell ribs, so that it was possible to continue the layer of boarding on top of the ribs over the beams. The angle of slope varies from roughly 45° at the tips of the cantilevered trusses to about 64° at the steel throat or funnel section. The geometries of the edge beams, developed from the geometry of the shell in cadwork, were exported via the DXF interface and manufactured on a 5-axis computerized numerical control (CNC) production line to a precision of ± 1 mm.

Modelling the centring

The centring ribs described a parabolic curve and, as a result of their 80-mm width, also rotated about their axis, which means that they had a constantly changing angle of inclination. Like the actual shell elements, the centring, from which the structure was ultimately removed, had to be constructed to a high degree of precision (tolerance ± 2 mm).

The grouting technique used in connecting the elements posed a special challenge in terms of the preparation of the work. As a result of the filling and ventilation holes required to ensure complete grouting, components that were otherwise identical when assembled in position had to be manufactured in completely different forms as a result of the different layout of the borings.

The preparatory work also included the simulation of the various situations for assembly. Care had to be taken that the assembly sequence did not lead to mutual obstruction between various components and that height limits were not exceeded so as not to impede the crane.

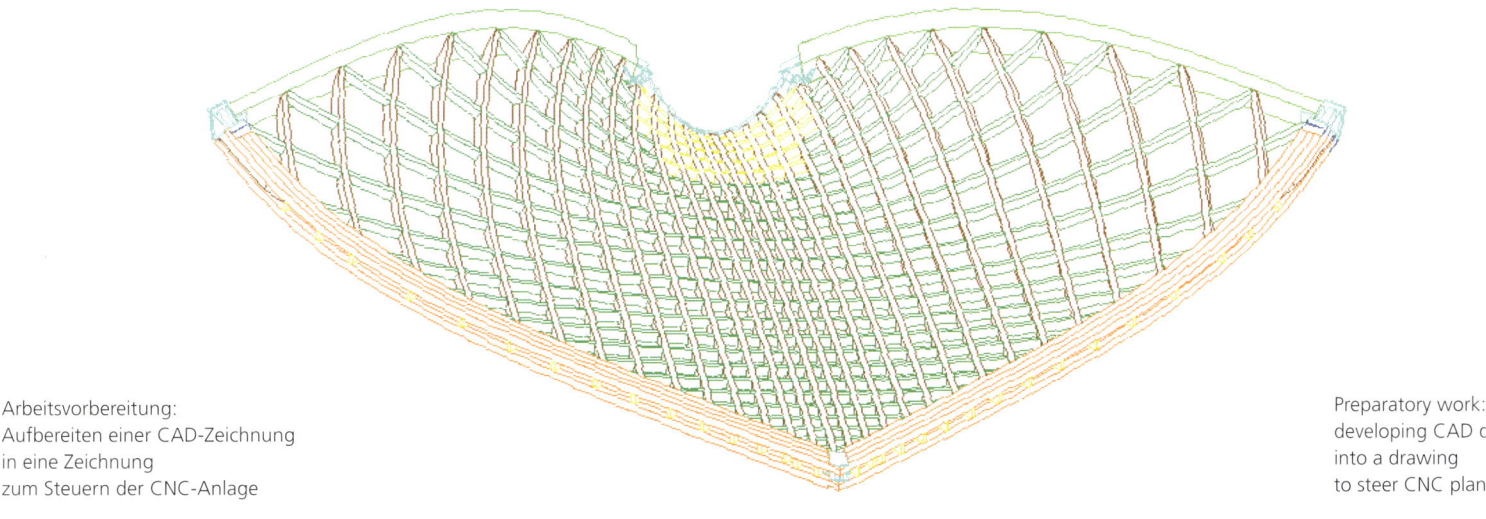

Arbeitsvorbereitung: Aufbereiten einer CAD-Zeichnung in eine Zeichnung zum Steuern der CNC-Anlage

Preparatory work: developing CAD drawing into a drawing to steer CNC plant

Transport und Montage

Peter Bertsche

Transport and Assembly

Peter Bertsche

Koppelung der Schirme

Connecting the Canopies

Peter Bertsche

Um beim Herstellen der Schirme zeitlich sowie örtlich möglichst unabhängig zu sein, erfolgte eine parallele Fertigung der verschiedenen Baugruppen der Schirme (Rippenschalen, Kragträger, Stahlpyramide und Turmkonstruktion) in verschiedenen Firmen. Die Anschlüsse an den Schnittstellen zwischen den Baugruppen mußten daher so entwickelt werden, daß eine möglichst einfache, paßgenaue und schnelle Montage möglich war.

Aus Gründen der höheren Paßgenauigkeit wurden mit Ausnahme der Verbindung von Schale und Randträger zum Kragträger alle Verbindungen als geschraubte Stahl-Stahl-Anschlüsse vorgesehen. Als Rahmen für die Fertigung und zum Stabilisieren bei Transport und Montage waren bereits beim Herstellen der Schale steife Randglieder erforderlich. Diese Funktion wurde von den Randträgern, den Kragträger-Untergurten (als Lastverteilungsträger) und dem gebogenen Kehlstahlteil erfüllt.

In der Planung wurde dabei das Ausheben der Schale aus dem Lehrgerüst, der Transport und das Einheben zwischen die Kragträger bedacht.

Es wurde daher eine stählerne Traversenkonstruktion vorgesehen, durch welche die Schalenform während des Montagezeitraums vom Ausheben der Schale bis zum Einhängen zwischen die Kragträger gewährleistet war. Die Traversen sind an insgesamt acht Punkten an die Randglieder der Schale angeschlossen. Die Anhängepunkte für den Kran befanden sich im Bereich der Kreuzungspunkte der großen Stahlprofile. Auch die Stahlpyramide wurde weitgehend im Werk vorgefertigt.

Auf Wunsch des Bauherrn wurde darüber hinaus eine Montagestatik erarbeitet, nachdem die beabsichtigten Transport- und Montagezustände mitgeteilt waren. Hierbei mußten von seiten des Ingenieurs folgende Zustände und Hilfskonstruktionen betrachtet und nachgewiesen werden:

- Traversenkonstruktion für die Rippenschalen,
- Traversenkonstruktion für den Tieflader,
- Hub- und Lagerungszustand für den Kragträger,
- vier Kragträger und eine Rippenschale montiert,
- vier Kragträger und zwei benachbarte Rippenschalen montiert,
- vier Kragträger und drei Rippenschalen montiert,
- Auflagerbock für die Abstützung des Kragträgers.

To remain as independent as possible of time and location in the erection of the canopies, the various structural sections (ribbed shells, cantilevered trusses, steel pyramids and towers) were constructed parallel to each other. An adequate fitting precision had to be ensured at the junctions between the various structural groups. The abutments had to be designed in such a way that the assembly could proceed as simply, precisely and quickly as possible.

To achieve greater precision, all junctions – with the exception of those between the shells and edge beams on the one hand and the cantilevered trusses on the other – were designed as bolted steel-to-steel connections. The rigid edge members that were necessary during the manufacture of the shells served as a supporting framework at the production stage and as a means of stabilizing the elements during transport and assembly. This function was performed by the edge beams, the lower chords of the cantilevered trusses (acting as load-distribution members) and the curved steel elements at the throat.

A steel spreader construction was, therefore, foreseen, which maintained the shell form during the period of assembly – from the removal of the elements from the centring to the time they were hoisted into position between the cantilevered trusses. The spreader construction was fixed to the edge members of the shell at eight points. Fixings for the attachment of crane cables were located near the points of intersection of the steel members. The steel pyramid was largely prefabricated at works as well.

At the wish of the client, structural calculations were also made for the erection process, once the transport and assembly conditions had been determined. The engineers had to take account of the following auxiliary structures and stages of construction:

- spreader construction for ribbed shells
- spreader construction for low-loader
- hoisting and storage conditions for the cantilevered trusses
- four cantilevered trusses and a ribbed shell assembled
- four cantilevered trusses and two adjoining ribbed shells assembled
- four cantilevered trusses and three ribbed shells assembled
- supporting construction for cantilevered trusses.

Die Konstruktion ist so bemessen, daß jeder Schirm für sich allein tragfähig und standsicher ist.

Das zu erwartende dynamische Verhalten – möglich ist vor allem die Torsionsschwingung der Konstruktion um die Turmachse und die Kippbewegung des Schirms um den schmalen Turmkopf – ließen jedoch eine Koppelung der einzelnen Schirme vorteilhaft erscheinen. Die durch die statisch mitwirkende Koppelung der Schirme entstehenden Koppelungskräfte, Schnittgrößen und Auflagerkräfte sind am Rechenmodell der gekoppelten Schirme berechnet. Die Auflagerkräfte sind im gekoppelten System gleichmäßiger verteilt. Die Extremwerte der Stützenbeanspruchung sind ebenfalls geringer.

Unter Schnee- und Windlasten betragen die maximalen rechnerischen Gesamtverformungen – einschließlich der Anteile aus der Verformung der Turmkonstruktion und der Stahlpyramide – an der Spitze des Kragträgers ca. 13 cm und an der freien Ecke der Schale ca. 36 cm. Die Durchbiegung unter Eigengewicht liegt bei 3,5 cm an der Kragträgerspitze und 7,3 cm an der freien Ecke. Eine maximale Verformungsdifferenz zwischen frei stehenden, nicht gekoppelten Schirmen würde etwa 17 cm an der Kragträgerspitze und etwa 50 cm an der freien Ecke der Schale betragen. Durch die Koppelung werden diese erheblichen Differenzen ausgeglichen.

The individual canopies that go to make up the structure are dimensioned in such a way that every one is a self-contained load-bearing unit and independently stable.

When subject to snow and wind loading, the maximum overall calculated deformation at the tips of the cantilevered trusses – including the element of deformation from the tower structure and steel pyramid – is approximately 13 cm, and at the free corners of the shells roughly 36 cm. The deflection caused by the dead load of the structure is around 3.5 cm at the tips of the trusses and 7.3 cm at the unsupported corners. A maximum difference of deformation between free-standing, non-coupled canopies would be roughly 17 cm at the tips of the trusses and about 50 cm at the unsupported corners of the shells. Torsional vibration about the axis of the tower and the overturning movement of the canopy about the narrow head of the tower were critical factors in the anticipated dynamic behaviour of the structure. In view of this, it seemed advantageous to connect the individual canopies to each other. The coupling forces, the permissible stresses and the support reactions resulting from the structural connection of the canopies were determined on the basis of a calculating model for the linked elements. In a coupled system, the reaction loads at the points of support are more evenly distributed. Extreme column-loading values are also reduced.

Unabhängige Bautechnische Prüfung

Independent Constructional Controls

Josef Lindemann, Martin Speich

Ausfahren der fertigen
Gitterschale aus
der Montagehalle
durch ein Spezialtor

Transporting the finished
lattice shell from
the assembly hall through
a special gate

◁◁ Blick von unten auf die
Koppelung von 4 Schirmen

◁◁ View from below
of connection between
the four canopies

Nach den Landesbauordnungen ist für Ingenieurbauwerke die Bautechnische Prüfung der Statischen Unterlagen durch die Untere Bauaufsichtsbehörde vorgesehen; diese schaltet bei Bauvorhaben höheren Schwierigkeitsgrades hierzu Prüfingenieure für das entsprechende Fachgebiet ein. Diese Überprüfung soll im Sinne des »Vier-Augen-Prinzips« durch eine unabhängige Instanz eine kritische Überprüfung der grundlegenden Annahmen, der rechnerischen Nachweise, der Konstruktionsdetails und der Bauausführung sicherstellen, vor allem im Hinblick auf Plausibilität, Tragfähigkeit und Gebrauchstauglichkeit und Übereinstimmung mit den allgemein anerkannten Regeln der Technik.

Prüfen der Planungsunterlagen in bezug auf Konstruktion und Berechnung des Tragwerks

Zunächst wurden in einer ersten Abschätzung Lastansätze nach DIN 1055 in Abstimmung zwischen Tragwerksplanern und Prüfingenieur formuliert. Es erwies sich nach ersten rechnerischen Abschätzungen als notwendig, für die Ermittlung der genaueren Verteilung der Wind- und Schneelasten Windkanalversuche durchführen zu lassen.

Bezüglich der Berechnungsansätze in der statischen Modellierung des Tragwerks mußten Näherungen in Hinsicht auf die Nachgiebigkeit des Verbundes der einzelnen Rippenelemente getroffen werden. Dies ist in Anlehnung an ein Verfahren gemäß DIN 1052 und auf der Basis von theoretischen Untersuchungen von Schelling geschehen. Mit Hilfe eines weiteren Verfahrens, das auf Überlegungen von Kreuzinger basiert und in die zukünftige DIN 1052 aufgenommen werden soll, wurden Vergleichsberechnungen durchgeführt. Zusätzlich wurden eigene numerische Modelle zum Überprüfen der getroffenen Annahmen verwendet. Besondere Überlegungen waren bei allen Berechnungsansätzen erforderlich, weil in den Kreuzungspunkten der einzelnen Schalenrippen jeweils nur eine Lamelle durchläuft, während die Lamelle der rechtwinklig hierzu verlaufenden Rippe gestoßen wird; dieses Prinzip wechselt in der nächsten Lamellenlage.

Obwohl alle zugrunde gelegten Ansätze unterschiedliche Ergebnisse liefern, war es möglich, auf diesem Wege eine hinreichend große Sicherheit bezüglich der Beurteilung des ermittelten Kraft- und Verformungszustandes der Schalenstruktur zu bekommen.

Ein Problem bei der numerischen Bearbeitung statisch hochgradig unbestimmter Systeme besteht darin, wirklichkeitsnahe Annahmen für die Steifigkeiten der Verbindungsmittel in den Anschlüssen zu treffen. Vergleichsberechnungen zeigen, daß bei Variation der Anschlußsteifigkeiten enorme Kräfteumlagerungen in der Schale stattfinden, so daß es erforderlich war, die Größenordnung der vorgelegten Berechnungsergebnisse durch Grenzwertbetrachtungen auf Plausibilität hin zu überprüfen.

Eine weitere Schwierigkeit bestand in einer ausreichend wirklichkeitsnahen Modellierung des Kragträgers und des Turms durch geeignete statische Systeme; auch hier war es erforderlich, im Rahmen der statischen Prüfung Vergleichsberechnungen mit jeweils veränderten Ansätzen durchzuführen, um so den Parameterbereich möglicher Beanspruchungen herauszufinden.

Da bei der vorliegenden Konstruktion die Tragwerkselemente relativ große Querschnittsabmessungen haben, war es erforderlich, jeweils genauere rechnerische Untersuchungen in den Lasteinleitungsbereichen der Randträger, der Kragträger und der Baumstämme durchzuführen, um insbesondere die auftretende Querzugbeanspruchung ausreichend genau abschätzen zu können. Dies gilt besonders auch für die Verbindung der Stämme untereinander, vor allen Dingen dort, wo die Aussteifungskonstruktion an eine Stammhälfte anschließt, gleichzeitig aber die gesamte Biegesteifigkeit des nachgiebig miteinander verbundenen zweiteiligen Querschnitts benötigt wird.

Besondere Sorgfalt war bei der Auswahl und der Güteüberwachung der halbierten Vollholzstämme erforderlich, da diese aus statischen Gründen einen ungewöhnlich großen Durchmesser besitzen. Für derartig voluminöse Konstruktionselemente aus Vollholz in einem Bauwerk sind bisher keine Erfahrungen bekannt, so daß eine besondere Qualitätskontrolle für diese Hölzer verlangt wurde. Auch für den Anschluß der neuartigen Verbindungselemente (BVD) zwischen den Baumstämmen und der anschließenden Stahlkonstruktion am Kopf- und Fußpunkt dieser Stämme wurden zusätzliche Versuche gefordert, die eine Aussage darüber liefern, wie sich die ansonsten bauaufsichtlich zugelassenen Elemente verhalten, wenn sie in frisches Holz eingebaut werden und nach Trocknung der Hölzer einer entsprechenden Belastung ausgesetzt werden. Hierzu wurden Bauteilversuche im Maßstab 1 : 1 durchgeführt.

Abweichungen von geregelten oder allgemein bauaufsichtlich zugelassenen Bauarten

Die Landesbauordnungen fordern, daß Baukonstruktionen durch geregelte Bauarten unter Verwendung von geregelten Bauprodukten (Materialien, Verbindungstechniken) gemäß Bauregelliste oder durch nicht geregelte Bauarten bei Vorliegen einer allgemeinen bauaufsichtlichen Zulassung (bzw. eines Prüfzeugnisses) erstellt werden. Sollen Tragwerkselemente abweichend hiervon hergestellt werden, ist eine Zustimmung im Einzelfall für diese Bauart durch die Obere Bauaufsichtsbehörde erforderlich.

Diese legt, normalerweise in Abstimmung mit der Unteren Bauaufsicht und dem Prüfingenieur, das weitere Vorgehen fest. In der Regel wird verlangt, daß zunächst einmal die Verantwortlichen (Planer, Tragwerksplaner, ausführende Bauunternehmung) eine Aussage darüber machen, welche Bauweise sie anwenden wollen und aufgrund welcher Gegebenheiten sie diese Ausführung für statisch ausreichend sicher halten. Zusätzlich wird in der Regel gefordert, daß ein ausgewiesener Spezialist – evtl. auf der Grundlage von Bauteilversuchen – eine gutachterliche Stellungnahme hierzu abgibt. Diese Unterlagen sind dann in Abstimmung mit der Unteren Bauaufsicht vom Prüfingenieur zu bewerten. Das Gesamtpaket wird danach der Oberen Bauaufsicht vorgelegt, die dann über eine Zustimmung entscheidet.

Bei der hier betrachteten Konstruktion handelt es sich um eine in vielerlei Hinsicht innovative Konstruktion, bei der insbesondere bezüglich der Verbindungstechniken die zuvor beschriebene Vorgehensweise prinzipiell erforderlich wurde war. Dies bezieht sich u. a. auf Blockverleimungen größerer BS-Holzquerschnitte im Bereich der Krag- und Randträger, auf flächige Verleimungen der Kerto-Furnierschichtholz-Stegplatten des Kragträgers und im Bereich der Kehle, auf Schraubpreßleimung einzelner Rippenlamellen vor Ort, und auf den Anschluß der Querrippen über schräg in das Hirnholz eingedrehte Schrauben.

Derartig schwierige Konstruktionen können natürlich nur mit Firmen realisiert werden, die über ausreichende Erfahrungen in vergleichbaren Techniken verfügen.

Überprüfen der Ausführung der Konstruktion

Beim Überprüfen vor Ort ist zunächst einmal die Qualität der verwendeten Materialien festzustellen. Insbesondere bei Leimarbeiten vor Ort ist stichprobenartig zu prüfen, ob die erforderlichen Randbedingungen eingehalten

werden. Im vorliegenden Fall war die Forschungs- und Materialprüfungsanstalt (FMPA) Stuttgart für die Verleimungen als qualitätsüberwachende Institution eingeschaltet. Mit der Qualitätssicherung der Vollholzstämme wurde das Labor für Holztechnik (LHT) Hildesheim beauftragt.

Besondere Nachweise waren u. a. auch für die durchzuführenden Schweißarbeiten und für die quer zur Blechdicke hoch beanspruchten Bauteile erforderlich.

Da bei einer derartigen Konstruktion die rechnerischen Schnittgrößen stark von der vorgegebenen Schalengeometrie abhängen, wurde die Schalengeometrie stichprobenartig in einem Lehrgerüst durch Nachmessen überprüft. Um eine Aussage über die Güte der verwendeten Berechnungsansätze zu erhalten, wurde gleichzeitig eine Überprüfung der Schalengeometrie im eingebauten Zustand unter Eigengewicht gefordert. Der Vergleich der rechnerischen und der gemessenen Verformung zeigt eine gute Übereinstimmung, was auf eine ausreichend gute Wahl der Ausgangsgrößen für die Modellierung der Tragstruktur schließen läßt.

Nach Vorlage des Transport- und Montagekonzeptes des ausführenden Unternehmens wurde schließlich für den Transport der Schalen vom Hub aus der Lehre bis zum Befestigen auf den Turm eine Montagestatik erforderlich, in der zu belegen war, daß die Schalen durch diesen Transportzustand an keiner Stelle so beansprucht werden, daß bezüglich der Standsicherheit bleibende nachteilige Auswirkungen auf den Endzustand zu erwarten sind (s. S. 60).

Für die Brandschutzbeurteilung der Konstruktion wurde ein Sondergutachter eingeschaltet.

Zusammenfassend ist hervorzuheben, daß bei diesem Bauwerk höchste Ansprüche von allen Beteiligten sowohl bei der Planung und als auch bei der Herstellung erfüllt werden mußten, insbesondere aufgrund des innovativen Charakters dieser herausragenden Holzkonstruktion.

State building regulations require all structural calculations for engineering structures to be submitted to the lower building supervisory authority for examination. In the case of projects of greater complexity, the authority will appoint proof engineers from the appropriate specialist field. The testing process is carried out by an independent expert or body of experts, based on the principle that two pairs of eyes are better than one. The process is intended as a critical examination of the basic assumptions, the calculations, the structural details and the method of executing of the work. In this respect, special attention is paid to the plausibility of the construction, its load-bearing capacity, functional appropriateness and observance of generally recognized rules of technology.

Examination of the planning documents in respect of construction and calculation

As a first step, the loading principles were formulated (based on German Standard DIN 1055) in a collaboration between the structural engineers and the proof engineer. Initial calculations showed that it would be necessary to conduct wind-tunnel tests to obtain more accurate data on the distribution of wind and snow loads.

In the calculating principles employed to create a statical model of the load-bearing structure, approximations of the compliant of the bonding of certain rib elements had to be made. This procedure was based on methods defined in German Standard DIN 1052 and on theoretical investigations carried out by Schelling. Comparative calculations were also compiled, using a method based on the theories of Kreuzinger (a method that is to be included in future versions of DIN 1052). In addition, a number of numerical models were used to check the assumptions that had been made. In determining the calculating principles, special consideration had to be given to the fact that at the points of intersection of the individual shell ribs, only single layers continue through, while alternate layers of the ribs at right angles to this are butt jointed at the sides.

Although these various lines of approach led to different results, it was nevertheless possible to obtain an adequate degree of certainty in judging the calculated forces and the deformation in the shell structure.

One basic problem with the numerical processing of statically highly indeterminate systems lies in the need to make relatively accurate assumptions for the rigidity of the connections at abutments. Comparative calculations show that with variations in the rigidity at the abutments, enormous redistributions of forces occur in the shell. It was necessary, therefore, to check the plausibility of the dimensions submitted with the calculated results by applying limiting values.

A further problem lay in creating a sufficiently accurate model of the cantilevered truss and the tower by means of appropriate static systems. Here, too, it was necessary, within the scope of the structural examination, to carry out comparative calculations based on various assumptions in order to determine the parameters of possible loading.

The load-bearing elements of the present structure have relatively large cross-sectional dimensions, so that a closer investigation of the zones where loads are transmitted to the edge beams, to the cantilevered trusses and to the tree stems was necessary. More precise calculations were required in this respect to make a sufficiently accurate estimate of the tension stresses per perpendicular to the grain. The same applies to the junctions between the individual stems, especially at the points where the bracing construction abuts a half-stem, but where, at the same time, the full bending stiffness of the compliantly connected two-part cross-section is required.

Special care had to be taken in the selection and quality control of the halved solid timber stems, which for structural reasons are of unusually large diameters. There are no known data on which to draw for the use in building of solid timber structural elements of such great volume. A special quality control was, therefore, required for these members. Additional tests were also demanded for the new BVD anchor elements between the tree stems and the steel construction at the head and foot. These tests provided information on how elements that are otherwise permitted in building construction behave when they are inserted in unseasoned timber and are then subject to the calculated loading after the timbers have dried. The tests were carried out on full-size constructional elements.

Montageablauf: Einheben einer Gitterschale bis zum fertigen Schirm /
Assembly sequence: from hoisting a lattice shell to its final fixing in the canopy

Zur Ökobilanz des EXPO-Daches

Life-Cycle Assessment of the EXPO Roof

Gerd Wegener, Bernhard Zimmer

Deviation from approved or permissible kinds of construction

State building regulations require that built structures in approved forms of construction should be constructed with recognized building products (materials, jointing techniques, etc.) in accordance with standard lists of approved materials and systems. Non-standard forms of construction require the general approval of the building supervisory authority (or a proof certificate). For structural elements that deviate from these conditions, it will be necessary to obtain a "permission in specific cases" for the relevant form of construction, granted by the upper building supervisory authority.

This authority will usually determine the further course of action to be taken in consultation with the lower building authority and the proof engineer. As a rule, the persons responsible for planning and executing the work (the planners, the structural engineers and the relevant construction companies) will be required to submit details describing the type of construction they wish to erect and evidence of its structural safety. In addition, a qualified specialist will normally have to submit a report on the subject, possibly on the basis of investigations carried out on specific constructional elements. The documents submitted will then be evaluated by the testing engineer in consultation with the lower building authority. Subsequently, the entire documents are forwarded to the upper building authority for a decision on whether permission can be granted or not.

In many respects, the roof under consideration here is an innovative form of construction. The special process for obtaining approval described above was necessary especially in respect of the jointing techniques. This applied to the following aspects: systems of block-gluing larger laminated timber cross-sections in the cantilevered trusses and edge beams; planar-gluing of the Kerto laminated timber web sheets in the cantilevered trusses and in the funnel or throat zone; screw-bonding and gluing of individual ribs on site; and connecting the cross-ribs by means of screws fixed on the splay in the end grain of the timbers.

Complex forms of construction of this kind can, of course, be realized only in collaboration with firms that have adequate experience in comparable techniques.

Supervising the execution of the structure

Controls carried out on site are meant to determine the quality of the materials used. Especially where structural members are glued on site, random checks should be made to ensure that the required conditions are observed. In the present scheme, the Research and Materials Testing Institute (FMPA), Stuttgart, was called in as a quality-control body to investigate the glue-bonded construction. The Laboratory for Timber Technology (LHT) in Hildesheim was entrusted with supervising the quality of the solid timber tree stems.

Special certificates were necessary for, among other things, the welding work and the constructional elements subject to heavy loading at right angles to the thickness of the metal.

The calculated forces for a structure of this kind are largely dependent on the geometry of the shell. This was, therefore, investigated by a process of measurement at random points after construction, but while still supported on the centring. To obtain some idea of the quality of the calculating methods applied, an examination of the shell geometry after fixing in position – under its own dead weight – was necessary. A comparison of the calculated and measured deformation reveals great correspondence between the two. One may, therefore, conclude that the dimensions assumed at the outset were appropriate for the design of the load-bearing structure.

After submitting the concept for the transport and assembly, compiled by the company responsible for the execution of the work, a structural calculation for the assembly was ordered, including the transport of the shells from the moment of raising them from the centring to the point when they were fixed to the tower. These calculations had to demonstrate that the shells would not be subject to stresses at any point in the process of transport and handling in a way that would have a lasting negative effect on their stability in their final state (see page 60).

A specialist consultant was also appointed to assess the structure in terms of fire protection.

In conclusion, it should be emphasized that, in view of the innovative character of the roof, this outstanding timber structure made exceptional demands of all those involved, both in the planning and in the manufacture and assembly.

Wer heute von Globalisierung spricht, denkt im allgemeinen an das Zusammenwachsen der Wirtschaftssysteme und der Märkte. Es findet aber seit mehreren Jahrzehnten auch eine Globalisierung der Umweltproblematik statt. Wurden in der Vergangenheit die lokalen und regionalen Umweltprobleme mit »End-of-pipe«-Technologien angegangen und »gelöst«, erfordern die heute erkennbar gewordenen globalen Umweltprobleme, wie der Anstieg des mittleren Temperaturniveaus und des Meeresspiegels sowie der Anstieg der CO_2-Konzentration in der Atmosphäre und die Folgen des anthropogenen Treibhauseffekts, ein generelles Umdenken sowie eine neue Qualität der Umweltpolitik und der Umweltvorsorge.

Das Thema der EXPO 2000 »Mensch – Natur – Technik« drückt die Herausforderung unserer Zeit aus, denn es beinhaltet alle die Zielgrößen der Nachhaltigkeit, wie sie die Staatengemeinschaft auf dem Weltklimagipfel in Rio de Janeiro 1992 formuliert hat: die Verbesserung der ökonomischen und sozialen Rahmenbedingungen bei gleichzeitiger langfristiger Sicherung der natürlichen Lebensgrundlagen.

Die Ansätze einer zeitgemäßen Umweltpolitik bzw. -vorsorge sind aber nicht gleichzusetzen mit einem Verzicht auf moderne Technologien, denn »High Tech« und Nachhaltigkeit schließen sich nicht generell aus. Sparsamer Umgang mit nicht erneuerbaren Ressourcen, hohe Effizienz bei der Energieumwandlung und niedriger Verbrauch von Energie, verbunden mit geringen Emissionen und der potentiellen Wiedereingliederung der Materialien in die natürlichen Kreisläufe der Natur am Ende der Nutzung, sind die Herausforderungen einer modernen Produktentwicklung.

Die Methode der produktbezogenen Ökobilanzierung ist eine der modernen Methoden, mit der die mit der Herstellung, der Nutzung und der Entsorgung eines Produktes verbundenen Umweltwirkungen erfaßt und bewertet werden können. Über den gesamten Lebensweg eines Produktes werden alle in das System ein- und ausfließenden Stoff- und Energieflüsse erfaßt, Umwelt- und Wirkungskategorien zugeordnet und anschließend bewertet. Dieser ganzheitliche, aber andererseits auch sehr komplexe Ansatz erlaubt beispielsweise den Vergleich von Produkten aus verschiedenen Materialien oder unterschiedlichen Produktionsweisen, und er soll Aufschluß geben über ihre »Umweltfreundlichkeit«.

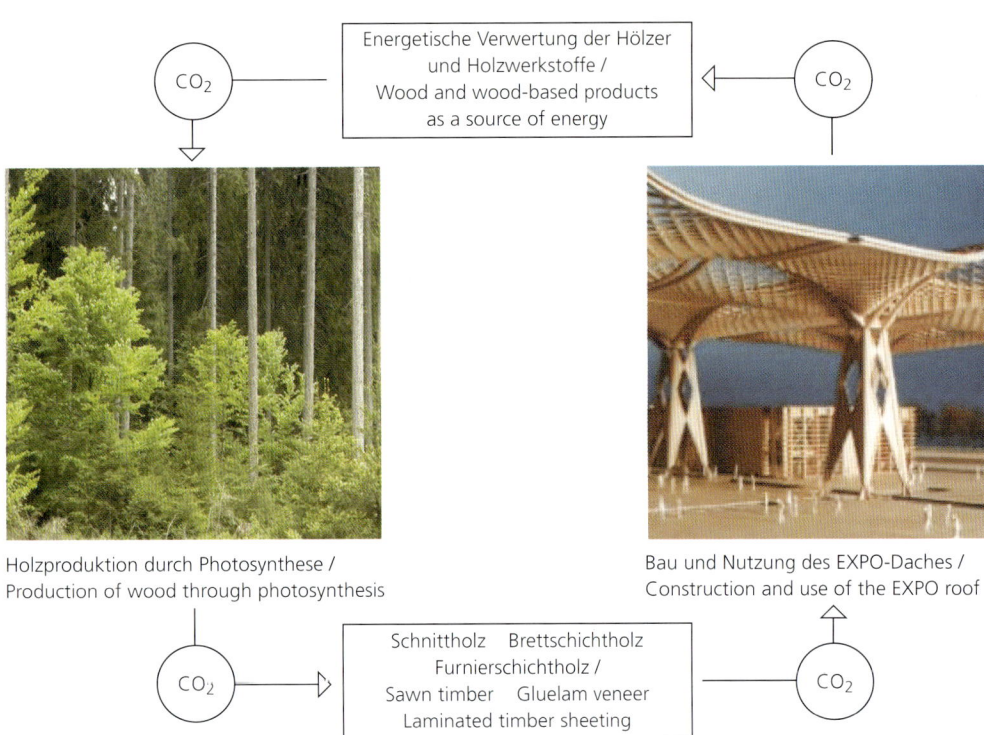

Kohlenstoff-Kreislauf: Verwertung des Holzes aus dem Wald über die Nutzung als Bauwerk bis zum Entsorgen / Carbon cycle: the utilization of wood from the forest as a building material up to its final disposal

Dem Bereich des Bauens kommt in diesem Zusammenhang in vielfacher Hinsicht besondere Bedeutung zu. Gebäude haben im Gegensatz zu vielen anderen Produkten eine sehr hohe Lebens- und Nutzungsdauer. Die Auswahl der für das Erstellen von Gebäuden zur Verfügung stehenden Materialien und Produkte und deren mögliche Kombinationen erscheinen nahezu unbegrenzt. Im Bauwesen treffen sich auch in besonderer Weise Ästhetik, Umwelt, technische Herausforderung sowie Ökonomie und menschliches Wohlbefinden. Architekten und Planer gestalten nicht nur Form und Gebrauchsfähigkeit, sondern sie verantworten auch die mit der Ausgestaltung und den Materialien verbundenen Wirkungen auf Mensch und Umwelt über die gesamte, oft mehrere Jahrhunderte dauernde Nutzungsdauer bis zur danach notwendigen Demontage und Entsorgung.

Besonders gilt dies für öffentliche und/oder besonders repräsentative Bauwerke, wie etwa das EXPO-Dach in Hannover. Architekten und Planer stehen, wenn sie das Anforderungsprofil erfüllen wollen, vor einer heute noch nahezu unlösbaren Aufgabe. Es ist eine bekannte Tatsache, daß die technischen Daten für neu entwickelte Baustoffe in der Regel zwar vorliegen, aber die ökologischen Kenndaten nur sehr lückenhaft sind, und es werden noch einige Jahre vergehen, bis es gesicherte Daten gibt. Das zeigt, daß bisher der Schwerpunkt im Bauwesen eindeutig im Bereich der Ästhetik und der technischen Machbarkeit lag.

Zukünftig – für die Neubauten auf dem Gelände der EXPO 2000 in Hannover hat diese Zukunft längst begonnen – wird es notwendig sein, ein Gleichgewicht zwischen den Einzelzielen zu finden um dem Ziel der Nachhaltigkeit möglichst nahe zu kommen.

Das EXPO-Dach ist auch in diesem Punkt ein gelungenes Beispiel. Es zeigt es einen Weg, wie zukunftsfähiges Bauen aussehen kann. Die Entwicklung während der Planung und Bauphase hat aber auch die Schwierigkeit aufgezeigt, die Einzelziele ins Gleichgewicht zu bringen. Vom ersten Entwurf, der Idee, der ersten Form und der Entscheidung für den Baustoff Holz war es die technische Umsetzbarkeit, die zur Herausforderung wurde. Zum Erreichen dieses Zieles mußte die Materialauswahl für dieses einzigartige Bauwerk angepaßt werden.

Holz ist der dominierende Baustoff, und damit erfüllt das EXPO-Dach, wie die bislang vorliegenden Ergebnisse zur Ökobilanzierung von Holz und Holzwerkstoffen zeigen, wesentliche Kriterien, die ein zukunftsfähiges Bauwerk erfüllen muß.

Holz ist immer auch Kohlenstoffspeicher. Mit der Holzproduktion durch die Bäume wird über die Photosynthese (siehe Abb. oben) CO_2 aus der Atmosphäre aufgenommen und in Holzsubstanz umgewandelt. Auf diese Weise bleibt der im Holz gebundene Kohlenstoff der Atmosphäre solange entzogen, wie das Holz nicht biologisch abgebaut oder beispielsweise über eine Verbrennung energetisch verwertet wird. Letzteres bietet die Möglichkeit, fossile Energieträger zu substituieren und damit nicht erneuerbare Ressourcen zu sparen.

Die Nutzung und Verwendung von Holz bedingt die Senkenwirkung unserer Wirtschaftswälder in Bezug auf CO_2. Anders als die Natur- und Primärwälder, die sich idealerweise in einer Art Fließgleichgewicht befinden, sind die nachhaltig bewirtschafteten Wälder durch die Entnahme von Holz immer wieder in die Lage versetzt, Biomasse zu akkumulieren. Durch die Holznutzung wird der Kohlenstoffspeicher dieser Wälder regelmäßig reduziert, es findet wieder eine Nettoproduktion an Holz und Biomasse auf der Fläche statt, und es wird dadurch mehr CO_2 gebunden als gleichzeitig durch biologischen Abbau freigesetzt wird.

Das System der Holznutzung und Holzverwendung hat weitere für die Umweltvorsorge günstige Eigenschaften. Das Gewinnen, das Verarbeiten und das Verwenden von Holz verbrauchen wenig Energie. Hinzu kommt, daß die benötigte Energie häufig aus Rest- oder Gebrauchshölzern, also aus regenerativen Energieträgern, gewonnen wird. Diese Doppelfunktion von Holz als Bau- und Werkstoff und als Energieträger ermöglicht eine nahezu perfekte Kreislaufwirtschaft, die schon heute als Modell für nachhaltiges Wirtschaften gelten kann.

Die Abbildung oben zeigt, daß sich auch beim EXPO-Dach nach der Nutzungsphase der Kohlenstoff-Kreislauf durch energetische Verwertung schließen läßt.

When people talk of globalization today, they are usually thinking of the coalescence of economic systems and markets. For several decades now, though, there has also been a globalization of environmental issues. In the past, local and regional problems were tackled and "resolved" with "end-of-the-line" technologies; in other words, not at the source. Today, however, environmental problems clearly have a global dimension. This can be seen in the increase in mean temperatures throughout the world, in the rise in the level of the oceans, in the increase in CO_2 concentrations in the atmosphere and in the outcome of the anthropogenic greenhouse effect. But it can also be seen in the fundamental process of rethinking that is taking place and in the new quality of environmental policies and the sense of care for the environment that are manifesting themselves.

The theme of EXPO 2000, "Humankind – Nature – Technology", is a succinct but unequivocal expression of the challenges facing our times; for it contains all the target issues associated with sustainability, as formulated by the community of nations at the World Climate Conference held in Rio de Janeiro in

1992: improving basic economic and social conditions, while at the same time maintaining the natural bases of subsistence in the long term.

The aims of modern environmental policies or environmental care cannot simply be equated with the abandonment of modern technology, however; for high technology and sustainability are not mutually exclusive per se. The challenges that have to be faced in developing modern products include the restrained use of finite resources, ensuring highly effective processes of energy conversion and low energy consumption, combined with low emissions and the potential reintegration of materials after use into the natural cycle.

Drawing up a life-cycle assessment for the manufacture, use and disposal of individual products is one of the modern techniques with which the environmental effects of a product can be assessed. All materials and energy flowing into and out of the system during the entire lifespan of a product are determined, assigned to environmental and impact categories and finally evaluated. This holistic but also extremely complex approach allows, for example, a comparison to be made between products consisting of various materials or manufactured by different production methods. It also affords an insight into the "environmental compatibility" of these products and production processes.

In this context, building plays a special role in several respects. In contrast to many other products, buildings have a great durability and a long life. The selection of materials and products and their possible combinations in the construction of a building seem almost unlimited. In the field of construction, however, a number of other factors also play an important role, in particular aesthetics, environmental considerations, technical challenges, economics and the sense of human well-being. Architects and planners design not only the form and functional efficiency of a building; through the fitting out and the choice of materials, they are also responsible for the effects it has on the people who use it and on the environment. These effects will make themselves felt over the entire lifespan of a building – possibly for hundreds of years – until it is subsequently dismantled or demolished and one has to dispose of the individual components.

This applies in particular to public buildings and/or structures that have a special formal presence, such as the EXPO roof in Hanover. Architects and planners are confronted today with an almost insoluble task if they wish to comply with the requirements listed above. It is a well-known fact that although the technical data for newly developed materials may be available, only very limited ecological benchmark figures are likely to exist; and many years will probably pass before any reliable data are available. This indicates that in building – and particularly in structures with a formal function – emphasis has clearly been placed hitherto on aesthetics and technical feasibility.

In the future – and for the new structures on the EXPO 2000 site in Hanover, this future began long ago – it will be necessary to seek a balance between all these constraints in order to achieve the overriding goal of sustainability to the fullest extent possible.

Here again, the EXPO roof may be seen as a successful model. Complying with the objectives of EXPO 2000, it shows how a viable form of building might look in the future. At the same time, developments during the planning and construction phases revealed the problems involved in trying to harmonize all the individual goals. From the initial design, the idea, the first form and the decision in favour of timber (our most important regenerable resource) as the material for the building, the technical implementation posed the greatest challenge. The choice of materials had to be adapted to the special needs of this unique structure.

Timber is the dominant material used here. The EXPO roof, therefore, fulfils the main criteria a building will have to meet in the future, as the life-cycle assessment for timber and timber products confirms.

Timber is not only the most important renewable building material and raw material in general. It is also a reservoir of carbon. As part of their growth, trees absorb CO_2 from the atmosphere and transform it by a process of photosynthesis into the substance of wood (see ill. p. 64). As a result, the carbon contained in the wood is removed from the atmosphere as long as the wood is not biologically degraded or used as an energy source, for example, through burning. (Wood, of course, represents a means of replacing fossil energy sources and thus conserving finite resources.)

The effect that commercial forests have in reducing CO_2 is dependent on the use of timber. In contrast to natural and primeval forests, which maintain an ideal steady state, forests that are subject to permanent commercial exploitation are continuously in a position to produce new biomass. Through the extraction and use of timber, the carbon reserves contained in these forests are constantly being depleted. A net production of wood and biomass occurs in these areas, with the result that more CO_2 is removed from the atmosphere than is released through biological degrading.

The use of timber has other beneficial effects for the environment, too, as is impressively demonstrated by the life-cycle assessments drawn up for wood and wood products to date. The extraction, processing and use of timber require relatively little energy. What is more, the energy used in the timber industries is often generated from offcuts or waste material; in other words, from renewable sources of energy. This dual function of wood – as a raw material and building product on the one hand, and as a source of energy on the other – facilitates an almost perfect cyclical economy, which can serve today as a model for sustainable forms of resource management.

The illustration on page 64 shows that in the case of the EXPO roof, at the end of its lifespan, the carbon cycle will be closed by exploiting the wood as a source of energy.

Daten zur Weißtanne

Data on the Silver Firs

Anzahl der eingeschlagenen Stämme	70
Größte Baumhöhe vor Abhieb	51 m
Größter Abhiebsdurchmesser	1,45 m
Mittlerer Durchmesser in Brusthöhe	1,37 m
Größter Zopfdurchmesser bei 18 Metern	90 cm
Alter	150 bis 300 Jahre
Stammlänge für Stütze	17,0 m
Volumeninhalt pro Stamm	9 bis 14 Festmeter
Gewicht pro Stamm nach Einschlag	8 bis 15 t
Auswahl und Einschlag im Wald	Dezember 1997 bis März 1998
Beginn der Lagerung	Ende April 99
Entrinden und Halbieren der Stämme	April / Mai 99
Ende der Lagerung	Anfang September 99
Anzahl der im Bauwerk enthaltenen Stämme	40
Mittlerer Zopfdurchmesser bei 17 m	74,1 cm ohne Rinde
Mittlerer Durchmesser Stammfuß	102,3 cm
Mittlerer Stamminhalt	10 Festmeter vor Abbund
Gewicht pro Stamm nach Trocknung	ca. 6 t
Länge für Einbau	16,10 m

Technische Prüfungen während der Lagerzeit
- Holzdichte anhand von Stammscheiben und Probewürfeln
- Holzdichte mit Ultraschallmessungen über die Gesamtlänge des Stammes
- Feuchteprofile für den Trocknungsverlauf
- Aststärken, Astansammlungen und -verteilungen
- Berechnungen hinsichtlich vorhandener Stammkrümmung und zu erwartender Auflasten für jeden Stamm
- Zusammenstellen von je vier sich ergänzenden Stämmen für einen Schirm

Ablauf der Vorarbeiten an den Stämmen
- Bearbeitungsdauer ca. 7 Wochen
- Biege- und Zugprüfungen von bis zu 4 m langen Stammteilen
- Einfräsen von Nuten, Beifräsen von Stammenden
- Einfräsen und Einbohren von Verankerungen mittels CNC-Maschinen
- Vormontieren von Stahlteilen
- Vormontieren von Beihölzern für Verbandscheiben
- Abhobeln von Astansätzen
- Transport nach Hannover

Vorteile von Weißtannenholz
- hohe natürliche Dauerhaftigkeit
- hohe Elastizität
- günstiges Schwundverhalten
- große Stammdimensionen verfügbar
- keine Harzgallen

Ökologische Eigenschaften der Weißtanne
- schattenertragend, ermöglicht damit den Aufbau stufiger Wälder (Plenterwälder)
- ermöglicht kahlschlagfreien Waldbau
- Wurzeln dringen tief in den Boden und sorgen für die nötige Stabilität der Wälder
- weitgehend resistent gegen Wind und Schneebruch

No. of trees felled	70
Max. tree height before felling	51 m
Max. cut diameter	1.45 m
Mean diameter at chest height	1.37 m
Max. diameter at head (at 18 m height)	90 cm
Age	150 to 300 years
Trunk length for columns	17.0 m
Volume of timber per trunk	9 to 14 m^3
Weight per trunk after felling	8 to 15 t
Selection and felling in forest	December 1997 to March 1998
Start of storage	end of April 1999
Removing bark and cutting trunks in half	April/May 1999
End of storage	beginning of September 1999
No. of tree trunks used in structure	40
Mean diameter at head (at 17 m length)	74.1 cm without bark
Mean diameter at foot	102.3 cm
Mean volume of trunk	10 m^3 before trimming
Weight per stem after seasoning	approx. 6 t
Length incorporated in structure	16.10 m

Technical tests during storage period
- density of wood based on slices from trunk and test cubes
- density of wood measured ultrasonically over entire length of trunk
- moisture curve during course of drying
- size, concentration and distribution of knots
- calculations of existing curvature of trunk and imposed loads expected for each trunk
- selection of groups of four complementary trunks for each canopy

Preparatory work on the tree trunks
- processing period: approx. 7 weeks
- bending and tension tests on trunk sections up to 4 m long
- cutting grooves; sinkings at ends of trunks
- cutting pockets and borings for anchorings with CNC-programmed equipment
- preassembly of steel components
- preassembly of ancillary timber elements for sheet bracing
- planing off projecting knots
- transport to Hanover

Advantages of silver fir timber
- great natural durability
- great elasticity
- favourable shrinkage behaviour
- large trunk diameters available
- no resin galls

Ecological properties of silver firs
- tolerate shade and thus facilitate the growth of mixed stands of varying height (suitable for selective felling)
- allow forestry in which clear-cutting can be avoided
- roots penetrate deep into the ground, ensuring requisite stability of forest
- robust and resistant to wind and snow damage

Daten zum Bauwerk

Data on the Structure

Zehn quadratische hölzerne Schirme aus je vier doppelt gekrümmten Rippenschalen mit einer Abmessung von je 40 x 40 m bilden ein Großdach über einer wettergeschützten Aktionsfläche. Sie ist durchzogen von einem System aus Wassergrachten. Auf den Standflächen befinden sich 4 Pavillons (Höhe: ca. 11 m, überbaute Fläche: 4800 m^2).

Ten square timber canopies roughly 40 x 40 m on plan – each consisting of four double-curved ribbed lattice shells – are joined together to form a large-area roof. The structure provides a covered outdoor space where a wide range of events can take place protected against the weather. The ground area is dissected by a system of water channels or "grachts". Beneath the roof, there are four pavilions roughly 11 m high and with a total footprint area of 4,800 m^2.

Dachfläche gesamt 16.000 m^2
(das entspricht der Größe von mehr als 2 Fußballfeldern)
Länge   160 m
Breite   120 m
Konstruktionshöhe 20 – 26 m
Wasserflächen (Grachten und Seefläche) ca. 15.000 m^2

Overall roof area 16,000 m^2
(larger than two football pitches)
Length 160 m
Width 120 m
Construction height 20-26 m
Area of water (channels and lake) ca. 15,000 m^2

Schirme		Abmessungen	39 x 39 m
		Abstände	1 m
		Relativbewegungen	+/- 6 cm
		Spielraum der Kopplungen	2 cm
Gründungen	10 Stück	Ringfundamente, Pfahlgründung, je 4 Pfähle ø 1,2 m, Tiefe: 10 bis 15 m	
Türme	10 Stück	Höhe: 17,6 m, Grundfläche: unten ca. 6 x 6 m, oben ca. 2,5 x 2,5 m	
Stützenfüße	40 Stück	Stahlbleche t: 20 bis 40 mm, Höhe: 1,3 m, Tiefe: 1,5 m	
Stützen	40 Stück	Weißtannen Vollholz, Einbaulänge: 16 m	
Turmaussteifung		Rahmenelemente mit Gurten aus Brettschichtholz und zweiseitiger Furnierschichtholz-Beplankung	
Stahlpyramiden	10 Stück	Höhe: 6,9 m, Stahlbleche t: 10 bis 60 mm, geschweißt	
Kragarme	40 Stück	Ober-/Untergurte, Diagonalen des K-Verbandes aus Brettschichtholz, Stegflächen und Aussteifungen aus Furnierschichtholz	
Rippenschalen	40 Stück	Brettstapelbauweise weitgehend verleimt, Rippen 8 bis 10 Lagen aus z. T. keilverzinkten Fichtenbrettern 16 x 3 cm	
Hochbelastete Zug- und Druckverbindungen		Vergußdübel: Schmiedestahl, Stabdübel, volumenneutraler Zementmörtel	
Dacheindeckung		Ethylen Tetra Fluor Ethylen-Film (ETFE-Film) für Flächen zwischen den Schirmen und auf den Kragträgern, voll recyclbar, Dicke: 0,2 mm, Transluzenz ca. 95 % Teflon-Glasfaser Membrane (PTFE-Glas), zusatzfreier Kunststoff ohne Weichmacher, wiederverwendbar Dicke: 0,9 mm, Transluzenz ca. 10 %	

Canopies		Dimensions	39 x 39 m
		Spaces between canopies	1 m
		Relative movements	± 6 cm
		Jointing tolerance	2 cm
Foundations	10 no.	ring foundations for towers each supported by four 1.20 m dia. piles, 10-15 m long	
Towers	10 no.	height: 17.6 m; dimensions at base: ca. 6 x 6 m; at top: ca. 2.5 x 2.5 m	
Column feet	40 no.	steel plates 20-40 mm thick; height: 1.30 m; depth: 1.50 m	
Columns	40 no.	solid silver fir trunks, constructional length: 16 m	
Tower bracing		framed elements consisting of laminated timber members with laminated wood sheeting on both faces	
Steel pyramids	10 no.	height: 6.90 m; welded steel plates 10-60 mm thick	
Cantilevered trusses	40 no.	upper and lower chords and diagonals of K-shaped bracing in laminated glued timber; webs and stiffening in laminated wood sheeting	
Ribbed lattice shells	40 no.	stacked-plank construction, largely glued; ribs with 8-10 layers of partly wedge-dovetailed softwood planks 16 x 3 cm	
Tension and compression connections subject to heavy loading		grouted dowels: wrought steel, dowels, grouting mortar not subject to autogenous volume change	
Roof covering		ethylenetetrafluoroethylene sheeting (ETFE) for areas between canopies and along cantilevered trusses: fully recyclable; thickness: 0.2 mm; translucence: ca. 95% Teflon and glass-fibre membrane (PTFE/glass-fibre) for canopy areas; without additives or plasticizers; reusable; thickness: 0.9 mm; translucence: ca. 10%	

Gewichte der einzelnen Baugruppen*		Massen für 10 Schirme*)	
Baumstamm für Stütze	ca. 6 t	Rundholz	ca. 500 m^3
Stütze (eingebaut)	ca. 10 t	Furnierschichtholz (FSH)	ca. 650 m^3
Kragarm	22 t	Brettschichtholz	ca. 1800 m^3
Gitterschale (Viertelschale)	37 t	Kantholz	ca. 150 m^3
Gitterschale inkl. Transporttraverse (Viertelschale)	46 t	Bretter	ca. 1950 m^3
Stahlpyramide	32 t	Baufurniersperrholz (BFU)	ca. 90 m^3
		Gesamt	ca. 5190 m^3

Weights of individual structural elements*		Volume of timber materials, 10 canopies*	
Tree-trunk column	ca. 6 t	Tree trunks	ca. 500 m^3
Column (built into structure)	ca. 10 t	Laminated timber sheeting	ca. 650 m^3
Cantilevered truss	22 t	Laminated timber members	ca. 1,800 m^3
Lattice shell (quarter shell)	37 t	Squared solid timber	ca. 150 m^3
Lattice shell (quarter shell) inc. spreader for transport	46 t	Boarding	ca. 1,950 m^3
Steel pyramid	32 t	Laminated construction board	ca. 90 m^3
		Total	ca. 5,190 m^3

* nach Angaben der Firma Merk-Holzbau

* Data supplied by Merk-Holzbau (timber construction company)

Szenen aus Planung und Werkstatt, vom Bau und von der Eröffnung

Scenes photographed during the planning and in the workshop, during construction and at the opening ceremony

Sepp D. Heckmann, Thomas Herzog (TH), Heinrich Cordes	Hanns Jörg Schrade, Roland Schneider, Rainar Herbertz
Dieter Kienast †, TH	TH mit dem Vorentwurfsmodell / with the preliminary design model
TH, Jacques-André Hertig	Heinrich Kreuzinger, Josef Lindemann, Martin Speich
Norbert Burger, Julius Natterer, Heinrich Kreuzinger, TH, Jaques-André Hertig	Ingo Brosch, TH
Martin H. Kessel, Bernhard Tritschler	Karl Moser, TH	Rainer Wittenborn

Julius Natterer, Roland Schneider, Jan Bunje, Michael Koch, TH	Roland Schneider, TH
Rudolf Grape † mit einem Schlossermeister / with a master-metalworker	Karl Moser, TH
Roland Schneider mit Mitarbeitern / with assistants	Zum Richtfest / At the topping-out ceremony
Peter Bertsche, Gerhard Ammann, Hanns Jörg Schrade	Sepp D. Heckmann, TH
Eröffnung / opening ceremony	Martin Pfundt	Verena Herzog-Loibl

Autoren des Buches und beteiligte Institutionen
entsprechend der Folge der Beiträge

Authors and Institutions Contributing to this Book
in the order of their contributions

Manfred Sack   Dr. phil., Dr.-Ing. E. h.
Von 1959 bis 1997 Feuilleton von ›DIE ZEIT‹, seitdem Freier Journalist. Schreibt über Architektur, Städtebau, Design und Unterhaltungskunst.

From 1959 to 1997: editorial contributor to arts section of the weekly newspaper 'DIE ZEIT'. Since 1997, freelance journalist. Writes on architecture, urban planning, design, and entertainment art.

Thomas Herzog   Univ.-Prof., Dr./Univ. Rom, Dipl.-Ing.
Herzog + Partner Architekten BDA. Wohn-, Gewerbe- und Ausstellungsbauten, Verwaltungsbauten, Bausysteme und -produkte unter Nutzung von Umweltenergien. Zahlreiche Preise und Auszeichnungen. Mitglied mehrerer Akademien der Künste. 10 Fachbücher.
Institut für Entwerfen und Bautechik, Lehrstuhl für Gebäudetechnologie der Technischen Universität München (TUM)

Herzog + Partner Architects BDA: housing, commercial, exhibition and administration buildings, construction systems and products exploiting environmentally sustainable forms of energy. Numerous prizes and awards. Member of several academies of art. Author and editor of 10 books on specialist subjects.

Thomas Kuckelkorn   Dipl.-Phys., Caroline Illinger   Dipl.-Phys.
– Tageslichtsimulation / Daylight simulation –
Institut für Entwerfen und Bautechik, Lehrstuhl für Gebäudetechnologie, TUM

Christoph Hoffmann   Dr. forest, Oberforstrat
Ministerium Ländlicher Raum, Ministerialrat Meinrad Joos, Landesforstverwaltung Baden-Württemberg (LFV), Stuttgart

Gerhard Rieger   Dr. rer. nat., Forstdirektor
Forstbetriebsgemeinschaft Kleines Wiesental,
Staatliche Forstämter Schopfheim und Todtmoos

Andreas Schabel   Forstrat
Forstdirektion Freiburg

Forstliche Versuchs- und Forschungsanstalt (FVA)
Abteilung Arbeitswirtschaft und Forstbenutzung
– Stammsortierung /Trunk grading and selection –
Ltd. FDir Dr. Mahler und Forstrat Wurster

Martin H. Kessel   Univ.-Prof., Dr.-Ing., Prüfingenieur für Baustatik
Beratender Ingenieur im Ingenieurbüro kgs
Institut für Baukonstruktion und Holzbau der Technischen Universität Braunschweig und Labor für Holztechnik der Fachhochschule (LHT), Hildesheim

Julius Natterer   Univ.-Prof., Dipl.-Ing.
Ingenieur- und Holzbau, Qualitätssicherungsmethoden, Verbindungs- und Verbundtechniken, Konstruktionen und Leichtbauweisen von Brettstapel- bis Holzflächentragwerken
Lehrstuhl und Institut für Holzkonstruktion (IBOIS)
an der Eidgenössischen Technischen Hochschule in Lausanne (EPFL)
Structural engineering and timber constructions, quality management, connectors and mixed structures, light weight structures in stacked-plank and shell constructions

Jean-Luc Sandoz   Prof. – Ultraschallmessungen / Ultrasonic measurements – EPFL

Klaus-Dieter Semmler   Dr. Mat. – Formfindung / Generation of forms –
Département de Mathematique Chaire de Géometrie, EPFL

Norbert Burger   Dr.-Ing., Alan Müller, Dipl.-Ing. EPFL
Johannes Natterer   Dipl.-Ing. EPFL
IEZ Natterer GmbH, Wiesenfelden
Internationales Entwicklungszentrum und Ingenieurbüro für Holzkonstruktionen

Jacques-André Hertig   Dr. ès Sc., Département de Génie Civil
Institut d'Hydraulique et d'Energie, EPFL

Heinrich Kreuzinger   Univ.-Prof., Dr.-Ing.
Institut für Tragwerksbau, Fachgebiet Holzbau, TUM

Robert Spengler   Dr.-Ing.
– Tragfähigkeitsuntersuchungen / Investigations of load-bearing capacity –
Institut für Tragwerksbau, Fachgebiet Holzbau, TUM

Werner Kelletshofer   Dipl.-Ing.
– Versuchsdurchführungen / Materials testing –
Materialprüfungsamt für das Bauwesen, Abteilung Holzbau, TUM

Peter Bertsche   Dipl.-Ing.
Ingenieurbüro Bertsche, Prackenbach
Entwicklung des Verbindungssystems Bertsche BVD;
Lösung entwicklungstechnischer Aufgaben /
Development of Bertsche BVD connection system; technical development solutions

Martin Pfundt   CAD + Dienstleistungen
Zimmermeister, Rheinfelden
Arbeitsvorbereitung mit CAD/CAM-Software, Ansteuerung von CNC-Abbundanlagen, Programmierung von Schnittstellen / Preliminary work, using CAD/CAM software; creation of data for operating cutting equipment; programming interfaces

Martin Speich   Prof. Dr.-Ing., Prüfingenieur für Baustatik
Josef Lindemann   Dipl.-Ing., Beratender Ingenieur
Ingenieurgemeinschaft Speich-Hinkes-Lindemann, Hannover
Büro für Tragwerksplanung

Forschungs- und Materialprüfungsanstalt für das Bauwesen (FMPA),
Otto-Graf-Institut, Universität Stuttgart
– Qualitätsüberwachende Institution für Verleimungen / Quality-control institution for glued bonding –

Gerd Wegener   Univ.-Prof., Dr., Dr. habil., Dr. h. c.
Bernhard Zimmer   Dr. rer. silv.
Institut für Holzforschung, TUM
Alle Bereiche der biologischen, chemischen und physikalisch-technologischen Holzforschung / All areas relating to biological, chemical and physical technology in timber research

Abendliche Stimmung mit leuchtenden Solarobjekten von Christoph Behling / Evening mood with shining solar objects by Christoph Behling

Förderer des Bauwerks

Sponsors of the Structure

Deutsche Bundesstiftung Umwelt, Osnabrück
– Forschung, Entwicklung und Optimierung /
  Research, development and optimization –

Bundesministerium für Bildung und Forschung (BMBF), Bonn
– Forschung und Entwicklung / Research and development –

Deutsche Forst- und Holzwirtschaft, Düsseldorf/Bonn
– Realisierung / Realization –   vertreten durch / represented by
- Bund Deutscher Zimmermeister, Berlin
- Gesamtverband Holzhandel BD Holz – VDH e.V., Wiesbaden
- Holzabsatzfonds, Bonn
- Landesforstverwaltungen
- Studiengemeinschaft Holzleimbau e.V., Düsseldorf
- Verband Niedersächsischer Zimmermeister, Hannover
- Vereinigung Deutscher Sägewerksverbände e.V., Wiesbaden
- VDMA – Fachverband Holzbearbeitungsmaschinen, Frankfurt(M)

Skyspan (Europe) GmbH, Rimsting/Chiemsee
– Membranhülle / Roof membrane –

Caparol GmbH & Co KG, Ober-Ramstadt
– Farbliche Fassung / Coloration –

EXPO 2000 GmbH, Hannover   – Infrastruktur / Infrastructure –

© 2000, Prestel Verlag, Munich · London · New York, and Thomas Herzog, Munich

Die Deutsche Bibliothek – CIP-Einheitsaufnahme
Ein Titeldatensatz für diese Publikation ist bei der Deutschen Bibliothek erhältlich
Die Deutsche Bibliothek – CIP Cataloguing-in-Publication-Data:
A catalogue record for this publication is available from Die Deutsche Bibliothek

Library of Congress Catalogue Card Number: 00-102993

Prestel Verlag, Mandlstrasse 26, D-80802 Munich, Germany
Phone +49 (089) 38 17 09-0, Fax +49 (089) 38 17 09-35;
175 Fifth Avenue, New York, NY 10010, USA
Phone +1 (212) 995-2720, Fax +1 (212) 995-2733;
Bloomsbury Place, London WC1A 2QA, UK
Phone +44 (020) 7323 5004, Fax +44 (020) 7636 8004

Organisation und wissenschaftliche Dokumentation /
Organization and documentation of research:
Sabine Djahanschah, Deutsche Bundesstiftung Umwelt;
Verena Herzog-Loibl, Mitarbeit / Assistant: Franziska von Wedel

Koordination und Buchgestaltung / Co-ordination and book design:
Verena Herzog-Loibl und / and Johannes Determann, Munich
Übersetzung / Translation: Peter Green, Munich / London
Reproduktion / Offset lithography: Medien Service Brodschelm, Munich
Satz / Composition: Max Vornehm GmbH, Munich
Druck und Bindung / Printing and binding: Aumüller Druck KG, Regensburg

Printed in Germany
ISBN 3-7913-2382-2

Zeichnungen / Drawings:
Herzog + Partner,
    Mitarbeit / Assistants: Peter Gotsch, Lavinia Herzog, Stefan Sinning:
    Seite / pages 19, 21, 22/23, 24, 26, 28/29, 30, 31, 32/33, 41, 43, 44, 45, 53, 54/55
IEZ Natterer GmbH: Seite / pages 42, 49, 51, 58
Martin Pfundt: Seite / pages 47, 58, 59

Photonachweis / Photo credits:
Aerophot Demuss: Seite / pages 40, 41
Ingo Brosch: Seite / pages 30 (3), 48 (2), 51, 60, 61, 63 (2), 70
Verena Herzog-Loibl: Seite / pages 12, 14, 23 (1), 30 (1), 34, 35, 36, 37, 38, 42 (2,3),
    44 (1), 45 (2,3), 48 (3), 49 (2), 55 (1, 2), 56, 59 (2), 62, 65, 67
Herzog + Partner: Seite / page 52, 68, 69
Jörg Koopmann: Seite / page 20
Moritz Korn: Seite / pages 21, 22, 23 (2,3), 42 (1), 43 (1,2), 45 (1), 57,
    58, 59 (1),
Labor für Holztechnik: Seite / page 39
Dieter Leistner: Umschlag / cover, Seite / pages 1, 2/3, 5, 6, 8/9, 10/11, 13, 27, 32,
    33, 44 (3), 46, 53, 66, 71, 72
Robertino Nikolic: Seite / page 43 (3)
Roland Schneider: Seite / pages 15, 30 (2), 31, 32 (2), 44 (2), 47, 48 (1), 49 (1), 60,
    63 (1)
Robert Spengler: Seite / page 56 (3)